Econometric Society Monographs No. 18

Two-sided matching

A study in game-theoretic modeling and analysis

Econometric Society Monographs

Editors:

Jean-Michel Grandmont   *Centre d'Études Prospectives d'Économie Mathématique Appliquées à la Planification, Paris*
Alain Monfort   *Institut National de la Statistique et des Études Économiques*

The Econometric Society is an international society for the advancement of economic theory in relation to statistics and mathematics. The Econometric Society Monograph Series is designed to promote the publication of original research contributions of high quality in mathematical economics and theoretical and applied econometrics.

Two-sided matching

A study in game-theoretic modeling and analysis

ALVIN E. ROTH
University of Pittsburgh

AND

MARILDA A. OLIVEIRA SOTOMAYOR
Pontificia Universidade Catolica do Rio de Janeiro

CAMBRIDGE
UNIVERSITY PRESS

Published by the Press Syndicate of the University of Cambridge
The Pitt Building, Trumpington Street, Cambridge CB2 1RP
40 West 20th Street, New York, NY 10011-4211, USA
10 Stamford Road, Oakleigh, Victoria 3166, Australia

First published 1990
First paperback edition 1992

Library of Congress Cataloging-in-Publication Data

Roth, Alvin E., 1951–
Two-sided matching : a study in game-theoretic modeling and
analysis / Alvin E. Roth and Marilda A. Oliveira Sotomayor.
p.   cm. – (Econometric Society monographs ; no. 18)
Includes bibliographical references.
ISBN 0-521-39015-X
1. Game theory.   2. Matching theory.   3. Econometric models.
I. Sotomayor, Marilda A. Oliveira.   II. Title.   III. Series.
HB144.R68   1990
519.3 – dc20   89-49114
                                                                CIP

British Library Cataloguing in Publication Data

Roth, Alvin E.   *1951–*
Two-sided matching.
1. Matching theory   2. Game theory
I. Title   II. Oliveira Sotomayor, Marilda   III. Series
511.64

ISBN 0-521-39015-X hardback
ISBN 0-521-43788-1 paperback

Transferred to digital printing 2003

TO EMILIE AND AARON
TO JORGE

Contents

Foreword

Robert Aumann

This book chronicles one of the outstanding success stories of the theory of games, a story in which the authors have played a major role: the theory and practice of matching markets. The theoretical part of the story begins in 1962, with the publication of the famous Gale–Shapley paper, "College Admissions and the Stability of Marriage." Since then, a large theoretical literature has grown from this paper, which is thoroughly covered in this book. But the most dramatic development came in 1984, when Roth published his discovery that the Gale–Shapley algorithm had in fact been in practical use already since 1951 for the assignment of interns to hospitals in the United States; it had evolved by a trial-and-error process that spanned more than half a century. The book describes this story in detail, as well as many other fascinating developments, on both the theoretical and practical side, and on the interface between the theoretical and the practical.

It is sometimes asserted that game theory is not "descriptive" of the "real world," that people don't really behave according to game-theoretic prescriptions. To back up such assertions, some workers have conducted experiments using poorly motivated subjects, subjects who do not understand what they are about and are paid off by pittances; as if such experiments represented the real world. We see here that in the *real* real world – when the chips are down, the payoff is not five dollars but a successful career, and people have time to understand the situation – the predictions of game theory fare quite well.

The authors are to be warmly congratulated for this fine piece of work, which is quite unique in the game-theoretic literature.

Acknowledgment

The work has been partially supported by grants from the National Science Foundation and the Alfred P. Sloan Foundation, and by the CNPq – the Conselho Nacional de Desenvolvimento Cientifico e Tecnologico, Brazil.

CHAPTER 1

Introduction

The purpose of this book is two-fold. First, it reviews and integrates the growing literature about a family of models of labor markets, auctions, and other economic environments. Second, we hope it will illustrate the subtle interactions between modeling considerations and mathematical analysis that characterize the use of game theory to explain and predict the behavior of complex "real-world" economic systems.

We will be concentrating on "two-sided matching markets." The term "two-sided" refers to the fact that agents in such markets belong, from the outset, to one of two disjoint sets – for example, firms or workers. This contrasts with commodity markets, in which the market price may determine whether an agent is a buyer or a seller. Thus whereas the market for gold has both sellers and buyers, any particular agent might be a buyer at one price and a seller at another, so the market is not two-sided in the sense we will speak of. But a labor market often is, since firms and workers are distinct. For example, as the wages of professors fall, some professors may leave the market, but none will become universities. The term "matching" refers to the bilateral nature of exchange in these markets – for example, if I work for some firm, then that firm employs me. This contrasts with markets for goods, in which someone may come to market with a truck full of wheat, and return home with a new tractor, even though the buyer of wheat doesn't sell tractors, and the seller of tractors didn't buy any wheat.

Although this book is not primarily concerned with reporting empirical observations about a variety of markets, it is intimately concerned with developing a theory capable of explaining such observations. Consequently the next two sections of this introduction will be devoted to a description of two sets of observations that will serve to motivate much of the material presented here. The first of these concerns the labor market that physicians in the United States face when they seek their first position following medical school. The second, very brief description, concerns a variety of auction markets, and concentrates on ways in which coalitions of agents can collude to influence the outcome. Auctions and entry-level labor markets are both two-sided matching markets in which many of the participants are in the market at the same time, and this feature makes them particularly amenable to analysis with the models developed here. One of the things we will see is that the incentives that any market organization gives to the participants impose constraints on what kinds of outcomes the market can achieve, whether it is organized in a centralized or decentralized manner. Much of our discussion will concern what can be accomplished subject to these constraints. In this sense much of this work is on the interface between economics and operations research.

Following those sections we will briefly discuss the nature of the game-theoretic tools that we will bring to bear on these phenomena, and describe the organization of the book.

1.1   The labor market for medical interns

1.1.1   *Some institutional history*

The following description is taken from Roth (1984a).

The internship was first introduced around the turn of the century as an optional form of postgraduate medical education. For students, internships offered a concentrated exposure to clinical medicine, and for hospitals they offered a supply of relatively cheap labor. The number of positions offered for interns was, from the beginning, greater than the number of graduating medical students applying for such positions, and there was considerable competition among hospitals for interns.

One form in which this competition manifested itself was that hospitals attempted to set the date at which they would finalize binding agreements with interns a little earlier than their principal competitors. As a result, the date by which most internships had been finalized began to creep forward from the end of the senior year of medical school. Many resolutions were passed and much moral suasion was applied in efforts to remedy this state of affairs, which was regarded as costly and inefficient both by the hospitals, who had to appoint interns without knowing their final grades or class standings, and by the students and medical schools, who found that schooling was disrupted by the process of seeking desirable appointments. Nevertheless, the advancement of the date of appointment continued at an accelerating pace through 1944, at which point it had reached the beginning of the junior year. Thus in 1944 the date of appointment had advanced to two full years before the internship would actually begin. At this point the Association of American Medical Colleges (AAMC) adopted a proposal that neither scholastic transcipts nor letters of reference would be released prior to the end of the junior year for students seeking internships commencing in 1946.

This proved to be an effective remedy for the problem it was intended to solve. Appointments for 1946 internships were largely made in the summer of 1945, and in subsequent years the dates at which information was released by medical schools was moved later into the senior year, and the date at which appointments were made followed in step. However a new problem appeared, and manifested itself in the waiting-period between the time offers of internships were first made, and the time students were required to accept them.

Basically, the problem was that students who were offered internships at, say, their third-choice hospital, and who were informed they were an alternate (i.e., on a waiting list) at their second choice, would be inclined to wait as long as possible before accepting the offered position in the hope of eventually being offered a preferable position. Students who were pressured into accepting offers before their alternate status was resolved were unhappy if they were ultimately offered a preferable position, and hospitals whose candidates waited until the last minute to reject them were unhappy if their preferred alternate candidates had in the meantime already accepted positions. Hospitals were unhappier still when candidates who had indicated acceptance subsequently failed to fulfill their commitment after receiving a preferable offer. In response to pressure originating chiefly from the hospitals, a series of small procedural adjustments were made in the years 1945–51. The nature of these adjustments, described next, makes clear how these problems were perceived by the parties involved.

For 1945, it was resolved that hospitals should allow students ten days after an offer had been made to consider whether to accept or reject it. In 1946, eight days were allowed. By 1949, the AAMC proposed that appointments should be made by telegram at 12:01 A.M. (on November 15), with applicants not being required to accept or reject them until 12:00 noon the same day. Even this twelve-hour waiting period was rejected by the American Hospital Association as too long. The joint resolution finally agreed upon contained the phrase "no specified waiting period after 12:01 A.M. is obligatory," and specifically noted that telegrams could be filed in advance for delivery precisely at 12:01 A.M. In 1950, the resolution again included a twelve-hour period for consideration, with the specific injunction that "hospitals and/or students shall not follow telegrams of offers of appointment with telephone calls" until the end of this period. (Note that the injunction against telephone calls was two-way, in order to stem a flood of calls both from hospitals seeking to pressure students into an immediate decision and from students seeking to convert their alternate status into a firm offer.)

By this time it was widely recognized both that there were serious problems in the last stage of the matching process, and that these problems could not adequately be resolved by compressing this last stage into a shorter and shorter time period. In order to avoid these problems and the costs they imposed, it was proposed and ultimately agreed that a more centralized matching procedure should be tried. Under this procedure, students and hospitals would continue to make contact and exchange information as before. (The complete job description offered by a hospital program in a given year was customarily specified in advance. Thus the responsibilities, salary, etc. associated with a given internship, even though they might be adjusted from year to year in response to a hospital's experience in the previous year's market, were not a subject of negotiation with individual job candidates.) Students would then rank in order of preference the hospital programs to which they had applied, hospitals would similarly rank their applicants, and all parties would submit these rankings to a central bureau, which would use this information to arrange a matching of students to hospitals, and inform the parties of the result. A specific algorithm was proposed to produce a matching from the submitted rankings.

It was agreed to try the proposed procedure for the 1950–1 market in a trial run that would *not* be used actually to match students and hospitals in that year. Instead, participants were asked to submit rankings "as if" they would be used for determining the final matching, and the plan would be evaluated for actual use after this trial run had occurred. On the basis of the trial run, the relevant medical associations agreed to adopt the procedure for the 1951–2 market. The procedure was to be employed on a voluntary basis: students and hospitals both were free either to participate in the process or to seek internship appointments on their own.

After this decision was announced, but before the procedure was implemented, objections by student representatives were raised to the algorithm used to produce a matching from the rankings submitted. Specifically, they observed that this algorithm made how adroitly a student composed his or her rank order list a matter of great importance. A student might suffer if he took a "flyer" and gave high rank to hospitals he preferred but had little chance of being matched with. That is, it was noted that a student who submitted a rank order of hospitals corresponding to his true preferences might receive a less preferable match than if he had submitted a different rank order. In response to these objections, a new algorithm was substituted for the old in the 1951-2 matching plan. This substitution was judged to be of sufficiently small import that its details were not widely disseminated, the announcements concerning the plan having already been distributed before the substitution was made. The substitute algorithm was used for the first time in 1951, and remains in use to this day. (This algorithm will be called the NIMP algorithm, where NIMP stands for National Intern Matching Program, which is the name under which the algorithm was initially administered. Today the program is called the National Resident Matching Program.)

Note that this system of arranging matches was conceived and implemented as a *voluntary* procedure – students and hospitals were free to try to arrange their own matches outside of the system, and there were no means of enforcing compliance on those who did participate. (The experience prior to 1950 amply demonstrated that no amount of moral suasion was effective at preventing participants from acting in what they perceived as their own best interests.) This makes it all the more remarkable that in the first years of operation, over 95% of eligible students and hospitals participated in the system, and these high rates of participation continued until the mid-1970's, around which time the rate of participation had dropped to about 85%. Increasing numbers of students, particularly those among the growing number of medical students married to other medical students, began to seek to arrange their own matches, without going through the centralized clearinghouse. The consequent disruption in the orderly operation of the market, reminiscent of the experience prior to the 1950's (although on a much smaller scale) has been a cause of concern among market participants.

Another source of concern about this market in the medical community has been the resulting distribution of physicians among hospitals. Rural hospitals get fewer interns than they wish, and a much higher percentage of interns who are graduates of foreign medical schools.

1.1.2   *The questions to be answered*

The advance of the date of appointment prior to 1945 and why the action of the AAMC was successful when previous actions had not been, is an important phenomenon that is paralleled in the history of a wide variety of labor markets. It can be understood in terms of relatively standard economic arguments about free-riding in the provision of public goods (with a late date of appointment being the public good). We will be concerned in this book with the less well understood phenomena that appeared in the market from 1945 on.

The chief phenomena we wish to explain are:

> What accounted for the disorderly operation of the market between 1945 and 1951?

> Why was the centralized procedure instituted in the 1951–2 market able to achieve such high rates of voluntary participation?

> Why did these high rates start to diminish by the 1970's, particularly among the growing number of medical students who were married to other medical students?

We will also want to investigate "strategic" questions of the kind that led to the scrapping of the "trial run" algorithm.

> Does the NIMP algorithm, as claimed by the sponsoring medical associations, give students and hospitals the incentive to submit rank orderings corresponding to their true preferences?

Finally, we would like to be able to get some idea of which aspects of the operation of the market could be influenced by modifying its organization, while preserving those features that have led to high rates of voluntary participation. In this regard, we will want to know:

> Can the defection of married couples be halted?

> Can the distribution of interns to rural hospitals be changed?

These points will be taken up in detail in Chapter 5, which studies a model of this market. The main ideas required for this explanation will first be developed by studying a simpler model in Chapters 2, 3, and 4. In order to help the reader keep in mind where we are heading, we turn now to a preview of the proposed explanation.

1.1.3   *A preview of the proposed explanation*

When we see that in the late 1940s there is a lot of two-way telephone traffic between hospitals and students, who sometimes renege on previous verbal agreements, we can hypothesize that there is some systematic incentive to the parties involved to behave in this way. These incentives must be mutual: If students who called hositals that had not extended them offers were uniformly told that no places were available, the practice would be unlikely to persist in the virulent form we have described. Situations in which there are some students and hospitals who are not matched with each other, but who *both* prefer to be matched one to the other, will therefore be called "unstable." By the same token, if the matching suggested by the NIMP algorithm was unstable in this way – that is, if there were students and hospitals that would prefer to be matched to one another rather than to accept the suggested match – then we would expect that these students and hospitals would continue to try to locate each other, and subsequently decline to accept the assignment suggested by the matching procedure. The very high rates of voluntary participation observed in the years following the introduction of the NIMP procedure suggest this was not the case and that the set of suggested assignments produced by the NIMP procedure must be "stable," in the sense of having the property that if some student would prefer another hospital to his or her suggested assignment, then that hospital does not return the favor, but instead prefers the students assigned to it rather than the student in question. In our mathematical analysis of the NIMP algorithm, we will see that its assignments do indeed have this property. So our explanation of why the chaotic market conditions prior to 1951 vanished following the introduction of the NIMP procedure will be that it introduced this kind of stability to the market.

In a similar vein, we will observe that as married couples became more common in this market, the procedures used to deal with them introduced instabilities once again, so married couples could find hospitals they preferred to their assigned matches and that were willing to offer them jobs. This will be the basis of our explanation of the defection of married couples from the system that became so noticeable in the mid-1970s. So the stability or instability of the system of producing matches will serve to help explain the market phenomena just discussed, as well as a number of other observations about this market that will be discussed in Chapter 5. (Chapter 5 will also briefly describe some observations from other, differently organized markets.) We will argue that the answers to our questions about how much freedom there is to alter the organization of the market while maintaining a high degree of voluntary participation also hinge on whether any given organization of the market leads to stable market outcomes.

A complementary set of ideas, having to do with the strategies of individual agents in the market, and with the kind of "strategic equilibrium" that might result, will be used to explore the question of whether, as claimed, it is always in the interest of all parties to state their true preferences. We will see that it is not, and that it cannot be for any procedure that produces stable outcomes. However it is possible to arrange things so that it is always in the best interest of *some* of the parties to state their true preferences. The development of these ideas will involve us in a number of subtle issues, not the least of which is that we will be forced to reconsider and reevaluate our conclusions about stability. If the students and hospitals may not be stating their true preferences when they submit rank-order lists for the NIMP algorithm, is there still reason to believe that the outcome is a stable set of assignments? It will turn out that there is.

1.2   Individual and collusive behavior in auction markets

We will again focus on strategic considerations when we turn to the study of auction markets, which, as we will see, can for many purposes be studied with the tools we will develop for labor markets. What strategic options are available to buyers and sellers at an auction obviously depends critically on the rules by which the auction is conducted, and there are a large number of different auction procedures that can be observed, used for different kinds of goods in different parts of the world. One commonly observed procedure is the so-called "English auction" (also called the open outcry ascending bid auction) in which the auctioneer solicits successively higher bids from the assembled bidders, who bid either by calling out their bids or (in another variation of the procedure) by indicating their acceptance of a bid called out by the auctioneer. The auction continues as long as some bidder is willing to raise the current bid. When no bidder is willing to do so the object being auctioned is awarded to the highest bidder, who pays the amount of his or her winning bid, provided that it is greater than a "reserve price" that may have been specified by the seller. If no bid higher than the reserve price is received, then the object is not sold, but reverts to the original owner. We will be particularly interested in the incentives that may exist for agents in an auction not to behave straightforwardly.

Not surprisingly, the opportunities to deviate profitably from straightforward behavior are different for buyers and sellers. The sellers (and their agent, the auctioneer) would like prices to be as high as possible, and the buyers would like prices to be as low as possible. The most commonly reported "strategic" behavior on the part of auctioneers or sellers is to introduce imaginary bids into the proceedings, which when practiced by auctioneers is called by a variety of colorful names, such as "pulling bids off the chandelier." (No less a painter than Rembrandt is reported to have bid on his own paintings at auction.) And the most commonly reported strategic behavior on the part of buyers is to form *rings* that agree to coordinate their bidding in an effort to keep down the price. Both practices are apparently pervasive, although various rules and regulations, some with the force of law, have been implemented or proposed in an effort to curb them.

In the summer of 1985, for example, the New York City Department of Consumer Affairs began an investigation of auction practices in the city, prompted by a lawsuit having to do with the well-known auction house Christie's, which in a 1981 auction had reported the successful sale of several paintings that had in fact not been sold, but had remained in the possession of the original owner. One of the regulations proposed by the department would have outlawed the practice of allowing sellers' reserve prices to remain secret, and so would have allowed bidders to ascertain more easily if the object being auctioned had in fact been sold when the auctioneer called a halt to the bidding. Other auction houses operating in New York joined Christie's in opposing such a measure, claiming it would inhibit their ability to deal with rings of bidders. For example, a spokesman for Christie's was quoted (Newsweek, 29 July 1985) as saying, "If we published the reserve, the dealers would form a ring, agree not to bid against each other, then have a mini-auction afterwards."

Such fears are not groundless, since bidder rings are widely reported to act in just such a way. For example, Cassady (1967) reports that in antique and art auctions, the subsequent auction among members of the ring, called a "knockout" auction, serves both to determine which of the ring members will receive what the ring has bought and what payments shall be made by ring members among themselves. (The *Oxford English Dictionary* cites nineteenth century sources for this meaning of the word *knockout,* suggesting that the organization of bidder rings in this way is not only a widespread phenomenon, but also not a new one.) Cassady remarks that buyer rings are common in many kinds of auctions all over the world, although in auctions of divisible commodities (such as fish in England, timber in the United States, and wool in Australia), rings commonly divide the purchase among themselves, rather than conducting a knockout auction. (The new regulations ultimately adopted in New York City continue to allow secret reserves, and allow auctioneers to introduce imaginary bids, but only up to the reserve price. Furthermore, the auctioneer must reveal that an object has not been sold before proceeding to the next lot [*New York Times,* 14 June 1987, p11].) An unusually detailed description of the strategic behavior of rings and auctioneers in New Jersey machine tool auctions is given by Graham and Marshall (1984).

The questions we would like to explore are these.

> What are the strategic possibilities open to auctioneers, and how are these related to the controversy over secret reserve prices?
>
> What are the strategic possibilities available to individual bidders and to rings of bidders?
>
> How is the kind of auction described here related to the kind of labor market described in Section 1.1?

In preparing to answer these questions, it will first be necessary to analyze how auctions work when no rings are present. For this purpose, the notion of stability introduced in our discussion of the hospital–intern labor market will once again prove useful. In an auction, the purchase of the object being auctioned at a given price by a given buyer is unstable if there is another buyer willing to pay a higher price. (The similarity with the hospital–intern market becomes apparent when we think of a sale as a matching of the seller with a buyer – at an unstable outcome, the seller could be matched with another buyer in a transaction that both parties would prefer.) We will see that under certain auction rules, a stable outcome will be achieved when the auctioneer and the bidders behave straightforwardly, and that, furthermore, individual bidders have no incentive to behave in any other way. But the auctioneer may profitably deviate from straightforward behavior, as may *coalitions* of bidders. We will see how the opportunities available to these coalitions of bidders might lead them to form rings roughly along the lines already described.

1.3   The game-theoretic approach

Game theory seeks to understand economic environments by analyzing how the motivations of the agents interact with the "rules of the game" – that is, the customs, rules, procedures, and constraints around which a market may be organized. In the course of our analysis, we will introduce a number of game-theoretic ideas as they apply to two-sided matching. We will occasionally digress to explain how the specific ideas we introduce are formulated in a more general game-theoretic context.

A subsidiary theme running through the material presented here is that the distinction commonly made between "cooperative" and "noncooperative" games is not a very clear one from the point of view of applications. As we will see, many economic environments can profitably be analyzed with the tools of *both* cooperative and noncooperative game theory. What is relevant to the choice of tools is often not so much the nature of the environment, but rather what kind of question is being asked. What primarily distinguishes the two kinds of theory is that the noncooperative theory works with relatively detailed models that specify the strategic choices available to individual agents, whereas the cooperative theory works with less detailed models that summarize the rules of the game in terms of what outcomes can be obtained by which coalitions of players.

In order to understand the difference between the approach taken in the tradition of cooperative game theory, and the more detailed approach taken in the tradition of noncooperative game theory, it may be useful to think of some other kinds of games. Suppose, for example, that we are interested in studying the workings of democracy. We might start by considering the consequences of the fact that many decisions in a democracy are made according to majority rule. The conclusions we could draw from this would presumably be applicable to many democracies, and would allow us to draw conclusions about what differences we might expect to find between democracies and, say, dictatorships. On a slightly more detailed level, we might note that legislative decisions in the United States are made by the actions of the two houses of Congress, together with the president. To become law, a bill needs to command a majority of the members of both houses, and a two-thirds majority is needed to pass a bill that the president opposes. The conclusions we would draw from this level of analysis should help us to better understand the workings of American democracy, and perhaps to understand how decision making under this system can be expected to differ from that in parliamentary democracies such as England or Israel.

But if we want to study the kinds of decisions American legislators have to make in the course of a session of Congress, we must examine the rules of legislation in much more detail. This is where we need to consider the rules of debate, which determine who has the right to bring issues to a vote, to propose amendments, and so forth. (The power of many important members of Congress derives not from the single votes they cast, but from their positions on important committees that are able to influence what legislation is brought to a vote, and when.) Only at this level of detail can we consider the strategic decisions that each legislator must make in pursuing his or her goals.

Of course, when we have considered the detailed rules, and how these influence the outcome, we need to check whether the results of a particular legislative process conform to the conclusions reached from the general, less detailed model of democracy. It could well be that some apparently democratic rules of debate and procedure might nevertheless effectively deliver all decision-making power into the hands of some minority. If so, our analysis at the detailed level would yield quite different conclusions than an analysis based on the presumption of majority rule.

Thus these more and less detailed approaches have complementary functions. The more abstract, less detailed models offer the possibility of yielding quite general conclusions, applicable to a variety of specific situations. The more detailed models allow us to reach stronger conclusions about specific situations, and to test the generality of the more abstract models. Of course, the conclusions reached about each kind of model also need to be evaluated by how well these conclusions can organize, explain, and predict empirical observations of the kinds of situations the models are meant to represent.

1.3.1   *The organization of this book*

This book is organized into three major parts. Part I deals with a simple two-sided matching model in which agents on each side of the market are matched with at most *one* agent on the other side. This model is therefore called the "marriage" model. Although it is too simple a model to permit us to draw conclusions about the kinds of labor market or auction phenomena described in Sections 1.1 and 1.2, many of the phenomena that we will discover when we investigate more complex models can be clearly seen in this simple model. Chapter 2 is devoted to introducing and developing the idea of stability for this model, and Chapter 3 looks further at the internal structure of the set of stable outcomes, and describes some computational algorithms. Chapter 4 considers questions of strategy, and how these relate to the discussion of stability.

Part II looks at more general models. Chapter 5 is devoted to an exploration of a model of the medical labor market already discussed. This model generalizes the marriage model in that one hospital may employ many students. We will see that although some of the conclusions of the marriage model do not carry over to the hospital–intern model, many do, and some of the striking conclusions about the marriage model become even more striking in the more general model. Since we will be concerned with explaining the historical development of that market, the chapter will be concerned not only with general properties of markets of this type, but will also analyze the particular matching algorithm implemented in that market in 1951. This model applies to many situations other than the hospital–intern market, however, and by way of indicating this, much of the discussion will be phrased in terms of "colleges" and "students," and could just as easily be phrased in terms of firms and workers. Other empirical studies will also be discussed. Chapter 6 considers further generalizations in which firms may have more complex, interdependent preferences over groups of workers and in which salary enters the model explicitly, rather than only implicitly.

Part III is further concerned with explicitly modeling monetary transfers. Whereas the models up to this point will have been discrete, the models in Part III treat money as a continuous variable. Chapter 7 explores a very simple model of auction markets, and specifically considers the kinds of auction behavior discussed in Section 1.2. Chapters 8 and 9 generalize this initial model in a number of dimensions. We will see that the continuous and discrete models, as well as those that deal with monetary transfers implicitly or explicitly, yield broadly similar results, although one or two significant differences will also emerge. One of the benefits of considering many different but closely related models is that it will give us an indication of the robustness of the various results we obtain. We will return to this point in the Epilogue.

In each of the following chapters, our general plan of attack is this: First we will introduce a formal model that describes the agents in the market, the possible outcomes, the preferences of the agents over the possible outcomes, and (at some appropriate level of detail) the rules that determine how the agents influence which of the possible outcomes will occur. Then we will propose a simple theory – in game theory, this is often called a *solution concept* – of what kinds of outcomes we can expect to observe. This theory will take into account the rules of the game and the preferences of the players. The body of the chapter will then be devoted to investigating mathematically the implications of the theory. These implications are what will ultimately allow us to test the theory against observed events in particular markets such as those discussed in Sections 1.1 and 1.2.

A note on the numbering.   Within each chapter, all numbered definitions, examples, theorems, propositions, lemmas and corollaries are numbered consecutively, so that Definition 2.1 may be followed by Example 2.2 and then Theorem 2.3. We hope this will make things easier to find when results in one part of the book are referred to in other parts.

1.4   Guide to the literature

Each chapter concludes with a "guide to the literature," in which we will indicate from where different results have been drawn, and remark on related topics of investigation that may not have been covered in the chapter.

The discussion of the hospital–intern labor market here comes from Roth (1984a), and some further issues concerning rural hospitals are discussed in Roth (1986). An early discussion in the medical literature of the algorithm adopted in that market can be found in Stalnaker (1953), and Checker (1973) notes the beginning of the decline in participation among married couples. A description and analysis of the many small regional markets for similar entry-level medical positions in England, Scotland, and Wales is contained in Roth (1989b). In spite of the differences between those markets and the American market described here, they all exhibit some striking similarities that underline the important role that stability, or the lack of it, plays in the behavior of the market.

Cassady (1967) provides a descriptive account of a large variety of auctions. Graham and Marshall's analysis of the auction behavior they observed can be found in their 1987 paper, which unfortunately does not contain as full a description of their empirical observations as does their original 1984 working paper.

PART I

One-to-one matching: the marriage model

In these chapters we will examine in detail the two-sided matching market without money that arises when each agent may be matched with (at most) one agent of the opposite set. For obvious reasons this model is, somewhat playfully, often called a "marriage market," with the two sets of agents being referred to as "men" and "women" instead of students and colleges, firms and workers, or physicians and hospitals. We will follow this practice here. In this whimsical vein, it may be helpful to think of the men and women as being the eligible marriage candidates in some small and isolated village.

The marriage market will be simpler to describe and investigate than a labor market in which a firm may employ many workers. And we will see in Part II that many (although not all) of the conclusions reached about this model will also apply to the hospital intern market, in which a hospital, of course, typically employs many interns. The marriage market will therefore be a good model with which to begin the mathematical investigation. In some of the discussion that follows, it will nevertheless be helpful to remember that much of our interest in this problem is motivated by labor markets, rather than by marriage in its full human complexity. (Thus we will sometimes speak about courtship, but never about dependent children or mid-life crises.)

In Chapter 2 we will study the class of stable outcomes to the marriage problem. In Chapter 3 the structure of the set of stable outcomes will be further studied, and computational algorithms will be presented. Chapter 4 will examine the strategic decisions that confront individual men and women, and the extent to which these can be affected by how the marriage market is organized. (Some readers may prefer to move directly from Chapter 2 to Chapter 4 after simply skimming Chapter 3.)

CHAPTER 2

Stable matchings

As in any game-theoretic analysis, it will be important in what follows to keep clearly in mind the "rules of the game" by which men and women may become married to one another, as these will influence every aspect of the analysis. (If, for example, our imaginary village were located in a country in which a young woman required the consent of her father before she could marry, then the fathers of eligible women would have a prominent role to play in the model.) We will suppose the general rules governing marriage are these: Any man and woman who both consent to marry one another may proceed to do so, and any man or woman is free to withhold his or her consent and remain single. We will consider more detailed descriptions of possible rules (concerning, e.g., how proposals are made, or whether a marriage broker plays a role) at various points in the discussion.

2.1 The formal (cooperative) model

The elements of the formal model are as follows. There are two finite and disjoint sets M and W: $M = \{m_1, m_2, \ldots, m_n\}$ is the set of men, and $W = \{w_1, w_2, \ldots, w_p\}$ is the set of women. Each man has preferences over the women, and each woman has preferences over the men. These preferences may be such that, say, a man m would prefer to remain single rather than be married to some woman w he doesn't care for.

A few words are in order about individuals' preferences, which play a critical role in game-theoretic models, and in economic models generally. An individual's preferences are meant to represent how he or she would choose among different alternatives, if he or she were faced with a choice. So when we say some individual *prefers* alternative a to alternative b, we mean that if that individual were faced with a choice between the two, he or she would choose a and not b, and if faced with a choice from a set of alternatives that included b, then he or she would not choose b if a were also available. When we say an individual is *indifferent* between the two, we mean that he or she might choose either one. We will say an individual *likes a at least as well as b* if he or she either prefers *a* to *b* or is indifferent between them.

To express these preferences concisely, the preferences of each man *m* will be represented by an ordered list of preferences, $P(m)$, on the set $W \cup \{m\}$. That is, a man *m*'s preferences might be of the form

$$P(m) = w_1, w_2, m, w_3, \ldots, w_p$$

indicating that his first choice is to be married to woman w_1, his second choice is to be married to woman w_2, and his third choice is to remain single. That is, if he is given the opportunity to choose to marry one of some set of women, then he will choose w_1 if she is one of the possibilities, but if neither w_1 nor w_2 is among the possibilities then he will choose to remain single. A man m' may be indifferent between several possible mates. This will be denoted by brackets in the preference list, so for example the list

$$P(m') = w_2, [w_1, w_7], m', w_3, \ldots, w_k$$

indicates that man m' prefers woman w_2 to w_1, but that he is indifferent between w_1 and w_7, and he prefers remaining single to marrying anyone else.

Similarly, each woman *w* in *W* has an ordered list of preferences, $P(w)$, on the set $M \cup \{w\}$. We will usually describe an agent's preferences by writing only the ordered set of people that the agent prefers to being single. Thus the preferences $P(m)$ just described will be abbreviated by

$$P(m) = w_1, w_2.$$

A little more terminology, and notation will prove useful. We will denote by *P* the set of preference lists $P = \{P(m_1), \ldots, P(m_n), P(w_1), \ldots, P(w_p)\}$, one for each man and woman. A specific marriage market will be denoted by the triple $(M, W; P)$. We write $w >_m w'$ to mean *m* prefers *w* to *w'*, and $w \geq_m w'$ to mean *m* likes *w* at least as well as *w'*. Similarly we write $m >_w m'$ and $m \geq_w m'$. Woman *w* is *acceptable* to man *m* if he likes her at least as well as remaining single, that is, if $w \geq_m m$. Analogously, *m* is *acceptable* to *w* if $m \geq_w w$. If an individual is not indifferent between any two acceptable alternatives, he or she has *strict preferences*.

Economists customarily make two assumptions about the preferences of an individual over any set *A* of alternatives. The first is that these preferences form a *complete ordering*. This means that any two alternatives can be compared – the individual may be indifferent between them, but he is never confronted with a choice he is unable to make. The second is that the preferences are *transitive,* which means that if *a* is liked at least as well as *b*, and if *b* is liked at least as well as *c*, then *a* is liked at least as well as *c*.

Individuals whose preferences possess these two properties are called *rational.* To get an informal idea of why this is not an abuse of the ordinary meaning of the word, consider a man who is not rational in this sense. Suppose, to make the example clear, that his preferences are not transitive, and that he prefers *a* to *b*, prefers *b* to *c*, and prefers *c* to *a*. Then you could presumably approach this person and engage in the following kind of profitable transaction. Suppose he is presently in possession of *a*. You offer to exchange it for *c* if he will give you a penny. Since he prefers *c* to *a* he accepts your offer. (Now he has *c*.) You now offer to exchange *c* for *b* if he will give you a penny. Again he accepts, and now he has *b*, and he is happy to pay you a penny in order to exchange it for *a*. He is now back where he started, minus three cents, and you have made a profit of three cents and are ready to operate this "money pump" again. Most of us would be prepared to say this individual has behaved irrationally. Although economists are well aware that the assumption that individuals always behave rationally is at best an approximation, it is nevertheless a useful one. All of the individual preferences discussed will be assumed to be complete and transitive. Consequently the preferences can be completely represented by preference lists, as discussed. (For much of our discussion, however, it would be sufficient to make only the weaker assumption that preferences are *acyclic,* that is, that cycles (e.g., $a >_m b >_m c >_m a$) of any length will not occur.)

Note that nothing about the economist's notion of rationality says anything about *what* an individual should prefer. The assumption of rationality is simply an assumption of certain sorts of regularities in an individual's choice behavior. An individual's preferences are our model of how he or she would choose among alternatives if faced with an individual choice. In the context of the marriage market, no single individual may choose a spouse, since a marriage requires the consent of both parties. The question posed by the marriage problem is therefore the following: Given the (individual) preferences of the many individuals involved, what kind of outcome will result from their (collective) interaction?

An outcome of the marriage market is a set of marriages. In general, not everyone may be married – some people may remain single. (We will adopt the convention that a person who is not married to someone is *self-matched*). Formally we have:

Definition 2.1. *A matching μ is a one-to-one correspondence from the set $M \cup W$ onto itself of order two (that is, $\mu^2(x) = x$) such that if $\mu(m) \neq m$*

then $\mu(m) \in W$ and if $\mu(w) \neq w$ then $\mu(w) \in M$. We refer to $\mu(x)$ as the mate of x.

Note that $\mu^2(x) = x$ means that if man m is matched to woman w (i.e., if $\mu(m) = w$), then woman w is matched to man m (i.e., $\mu(w) = m$). The definition also requires that individuals who are not single be matched with agents of the opposite set – that is, men are matched with women. These two requirements explain why matchings can be thought of as sets of marriages.

A matching will sometimes be represented as a set of matched pairs. Thus, for example, the matching

$$\mu = \begin{matrix} w_4 & w_1 & w_2 & w_3 & (m_5) \\ m_1 & m_2 & m_3 & m_4 & m_5 \end{matrix}.$$

has m_1 married to w_4 and m_5 remaining single; that is, $\mu(m_1) = w_4$ and $\mu(m_5) = m_5$, and so on. We will present the pairs in a matching either in the order of the men or of the women, so that the above matching could also be represented as

$$\mu = \begin{matrix} w_1 & w_2 & w_3 & w_4 & (m_5) \\ m_2 & m_3 & m_4 & m_1 & m_5 \end{matrix}.$$

Each agent's preferences over alternative matchings correspond exactly to his or her preferences over his or her own mates at the two matchings. Thus man m, say, prefers matching μ to matching ν if and only if he prefers $\mu(m)$ to $\nu(m)$. Thus we are assuming that man m cares about who *he* is matched with, but is not otherwise concerned with the mates of other agents.

2.2   Stable matchings – a theoretical framework

We can now start to consider the elements of a theory about which matchings are likely to occur, and which are not. This is where the rules of the game play a critical role. The first element of our theory is that since the rules specify that no agent may be compelled to marry, we will not observe any matchings that could only result from compulsion of one of the agents. Specifically, consider a matching μ that matches a pair (m, w) who are not mutually acceptable. Then at least one of the individuals m and w would prefer to be single rather than be matched to the other. Such a matching μ will be said to be *blocked* by the unhappy individual. (In the terminology of game theory, such a matching is said to be *individually irrational*.) The motivation for this terminology should be clear. If man

m, say, prefers remaining single to marrying woman w, this means that if he were faced with a choice, he would never choose to marry woman w so long as remaining single was one of his alternatives. But the rules of the game insure that the option of remaining single is always available. So the matching μ will not occur, since it would require man m's consent, and he will demur in favor of remaining single.

So the first element of our theory is that individually irrational matchings will not occur, which is to say that only matchings that do not compel individuals into unacceptable matches will occur. These are identified by the following definition.

Definition 2.2. *The matching μ is **individually rational** if each agent is acceptable to his or her mate. That is, a matching is individually rational if it is not blocked by any (individual) agent.*

It is clear that, no matter what preferences the agents have, at least one individually rational matching will exist, since the matching that leaves every agent single is always individually rational. However a theory that predicts nothing more than that an individually rational matching will occur is not likely to tell us much. To obtain stronger predictions, we will have to go beyond simple considerations of individual rationality, and begin to consider what we might be able to say about agreements between individuals.

Consider a matching μ such that there exist a man m and a woman w who are not matched to one another at μ, but who prefer each other to their assignments at μ. That is, suppose that $w >_m \mu(m)$ and $m >_w \mu(w)$. The man and woman (m, w) will be said to *block* the matching μ. Again, the motivation should be clear. Suppose such a matching μ should be under consideration – suppose, for the sake of clarity, that no agreements have been reached yet, but that courtships are under way that, if concluded successfully, will result in the matching μ. This state of affairs would be unstable in the sense that man m and woman w would have good reason to disrupt it in order to marry each other, and the rules of the game allow them to do so. So we now have two criteria for excluding potential matchings from our theory of which matchings can be expected to occur. The matchings that remain are those that meet the following definition.

Definition 2.3. *A matching μ is **stable** if it is not blocked by any individual or any pair of agents.*

The following example should help make the definition clear.

Example 2.4

There are three men and three women, with the following preferences.

$$P(m_1) = w_2, w_1, w_3 \qquad P(w_1) = m_1, m_3, m_2$$
$$P(m_2) = w_1, w_3, w_2 \qquad P(w_2) = m_3, m_1, m_2$$
$$P(m_3) = w_1, w_2, w_3 \qquad P(w_3) = m_1, m_3, m_2$$

All possible matchings are individually rational (since all pairs (m, w) are mutually acceptable). The matching μ given by:

$$\mu = \begin{matrix} w_1 & w_2 & w_3 \\ m_1 & m_2 & m_3 \end{matrix}$$

is unstable, since (m_1, w_2) is a blocking pair. However the matching

$$\mu' = \begin{matrix} w_1 & w_2 & w_3 \\ m_1 & m_3 & m_2 \end{matrix}$$

is stable.

We now have some building blocks for a simple theory, which we might state (a little too simply) as "Only stable matchings will occur." Of course, as with most simple theories, we should keep in mind that we will want to attach some qualifications to it when we examine specific situations. For example, in a marriage market in which men and women don't know each others' preferences, an unstable matching μ that is blocked by some pair (m, w) might occur simply because m and w weren't aware of each other's interest. And unstable matchings might occur in a market whose rules for proposing marriages don't make it easy for blocking pairs to form. For the moment, though, let us keep in mind marriage markets in which the agents have a very good idea of one anothers' preferences, and have easy access to each other. In such markets, we might expect that stable matchings will be especially likely to occur. By examining their properties from a number of points of view, we can therefore hope to gain insight into these markets.

Another context in which stable matchings might be particularly important is in the hypothetical problem facing an entrepreneurial matchmaker, eager to set up business in a small town in which marriages have previously been arranged without benefit of such services. (Note the parallel here with the situation existing around 1951 in the medical labor market discussed in Section 1.1.) This matchmaker will be a success only if couples take her advice, and they are not compelled to do so. If the matchmaker proposes a set of marriages μ that form an unstable matching, then the men and women who form blocking pairs will find that they can do better by ignoring the matchmaker and marrying each other. However if the matchmaker proposes a stable matching μ, then any individual who is unhappy with his or her proposed match finds that all the spouses he or she would prefer are not interested in eloping, but would instead prefer to go through with the matchmaker's plan. (Note again how our concern with this definition of stability reflects the rules of the game. If couples were compelled to take the matchmaker's advice, the need to propose stable matchings would disappear.) So in this context also, we might expect that stable matchings will be observed.

2.3   Some properties of stable matchings

Before we can proceed very far in exploring this kind of theory, we need to ask a fundamental question: Do stable matchings always exist? We have looked at one example that tells us that stable matchings *can* exist, but it might be that in some other example, every matching was unstable. That would call for a major revision of any theory that suggests that unstable matchings are unlikely to occur.

We will demonstrate, however, that this problem does not arise. For any marriage market, no matter how many men and women are involved, and no matter what preferences they have, there will always be at least one matching that is stable. This is quite a surprising result, which turns out to depend on both the two-sidedness of the market and the fact that matchings are one-to-one, as the following three examples make clear. Example 2.5 concerns matching among agents of a single kind, and Example 2.6 concerns matching among three kinds of agents. Example 2.7 concerns matching that is not one-to-one. For each of these problems, it is straightforward to define a notion of stability corresponding to what we have defined for the marriage problem, but the examples show in each case that the preferences of the agents may be such that no stable matchings exist.

Example 2.5: The roommate problem (Gale and Shapley)
There is a single set of n people who can be matched in pairs (to be roommates in a college dormitory, or partners in paddling a canoe). Each person in the set ranks the $n-1$ others in order of preference. An outcome is a matching, which is a partition of the people into pairs. (To keep things simple, suppose the number n of people is even.) A *stable* matching is a matching such that no two persons who are not roommates both prefer each other to their actual partners.

Consider four people: a, b, c, and d, with the following preferences:

$$P(a) = b, c, d$$
$$P(b) = c, a, d$$
$$P(c) = a, b, d$$
$$P(d) = \text{arbitrary.}$$

Person d is the last choice of everyone else. (Perhaps you know someone like that.) Each of the other people is someone else's *first* choice. So no matching will be stable, since any matching must pair someone with agent d, and that someone will be able to find another person to make a blocking pair. That is, the possible matchings are

$$\mu_1 = \begin{matrix} c & a \\ b & d \end{matrix}, \quad \mu_2 = \begin{matrix} a & d \\ b & c \end{matrix}, \quad \mu_3 = \begin{matrix} b & a \\ d & c \end{matrix}.$$

And (c, a), (b, c), and (a, b) block μ_1, μ_2 and μ_3, respectively.

Example 2.6: The man-woman-child marriage problem
(Alkan)
There are three sets of people: men, women, and children. A matching is a division of the people into groups of three, containing one man, one woman, and one child. Each person has preferences over the sets of pairs he or she might possibly be matched with. A man, woman, and child (m, w, c) *block* a matching μ if m prefers (w, c) to $\mu(m)$; w prefers (m, c) to $\mu(w)$, and c prefers (m, w) to $\mu(c)$. A matching is stable only if it is not blocked by any such three agents.

Consider two men, two women, and two children, with the following preferences:

$$P(m_1) = (w_1, c_3), (w_2, c_3), (w_1, c_1), \dots \text{(arbitrary)}$$
$$P(m_2) = (w_2, c_3), (w_2, c_2), (w_3, c_3), \dots \text{(arbitrary)}$$
$$P(m_3) = (w_3, c_3), \dots \text{(arbitrary)}$$
$$P(w_1) = (m_1, c_1), \dots \text{(arbitrary)}$$
$$P(w_2) = (m_2, c_3), (m_1, c_3), (m_2, c_2), \dots \text{(arbitrary)}$$
$$P(w_3) = (m_2, c_3), (m_3, c_3), \dots \text{(arbitrary)}$$
$$P(c_1) = (m_1, w_1), \dots \text{(arbitrary)}$$
$$P(c_2) = (m_2, w_2), \dots \text{(arbitrary)}$$
$$P(c_3) = (m_1, w_3), (m_2, w_3), (m_1, w_2), (m_3, w_3), \dots \text{(arbitrary)}.$$

There is no stable matching in this example. In fact,

1. All matchings that give m_1 (respectively m_2 and w_2) a better family than (m_1, w_1, c_1) (respectively (m_2, w_2, c_2)) are unstable.

To see this, note that any matching containing either (m_1, w_1, c_3) or (m_2, w_2, c_3) is blocked by (m_3, w_3, c_3) and any matching containing (m_1, w_2, c_3) is blocked by (m_2, w_3, c_3).

2. Any matching that does not contain (m_1, w_1, c_1) (respectively (m_2, w_2, c_2)) is either blocked by (m_1, w_1, c_1) (respectively (m_2, w_2, c_2)) or is unstable as already shown in item 1 above.
3. Finally, (m_1, w_2, c_3) blocks any matching that contains (m_1, w_1, c_1) and (m_2, w_2, c_2). So all matchings are unstable.

Observe that the preferences in this example are "separable" into preferences over men, women, and children; that is, there are no preferences such that, e.g. (m, w, c) is preferred by m to (m, w, c'), but (m, w', c') is preferred to (m, w', c).

Example 2.7: Many-to-one matching
Consider a set of firms and a set of workers. Each worker can work for at most one firm and has preferences over those firms he or she is willing to work for. Each firm can hire as many workers as it wishes and has preferences over those subsets of workers it is willing to employ. It is clear what a matching is in this case, and a firm F and a subset of workers C *block* a matching μ if F prefers C to the set of workers assigned to it at μ, and every worker in C who is not assigned to F prefers F to the firm he or she is assigned by μ. Consider two firms and three workers with the following preferences:

$$P(F_1) = \{w_1, w_3\}, \{w_1, w_2\}, \{w_2, w_3\}, \{w_1\}, \{w_2\}$$
$$P(F_2) = \{w_1, w_3\}, \{w_2, w_3\}, \{w_1, w_2\}, \{w_3\}, \{w_1\}, \{w_2\}$$
$$P(w_1) = F_2, F_1$$
$$P(w_2) = F_2, F_1$$
$$P(w_3) = F_1, F_2.$$

The only individually rational matchings without unemployment are:

$$\mu_1 = \begin{matrix} F_1 & F_2 \\ \{w_1, w_3\} & \{w_2\} \end{matrix}, \text{ which is blocked by } (F_2, w_1)$$

$$\mu_2 = \begin{matrix} F_1 & F_2 \\ \{w_1, w_2\} & \{w_3\} \end{matrix}, \text{ which is blocked by } (F_2, \{w_1, w_3\})$$

$$\mu_3 = \begin{matrix} F_1 & F_2 \\ \{w_2, w_3\} & \{w_1\} \end{matrix}, \text{ which is blocked by } (F_2, \{w_1, w_2\})$$

$$\mu_4 = \begin{matrix} F_1 & F_2 \\ \{w_2\} & \{w_1, w_3\} \end{matrix} \quad , \quad \text{which is blocked by } (F_1, \{w_2, w_3\})$$

$$\mu_5 = \begin{matrix} F_1 & F_2 \\ \{w_1\} & \{w_2, w_3\} \end{matrix} \quad , \quad \text{which is blocked by } (F_2, \{w_1, w_3\}).$$

Now observe that any matching that leaves w_1 unmatched is blocked either by (F_1, w_1) or by (F_2, w_1); any matching that leaves w_2 unmatched is blocked either by (F_1, w_2), (F_2, w_2), or $(F_2, \{w_2, w_3\})$. Finally, any matching that leaves w_3 unmatched is blocked by $(F_2, \{w_1, w_3\})$.

Returning now to the marriage problem, a natural way to prove that stable matchings always exist would be to demonstrate some procedure or algorithm that, when applied to any marriage problem, would produce a stable matching.

One procedure that might suggest itself is to start with an arbitrary matching, and if it is blocked by, say, man m and woman w, make a new matching in which m and w are matched to one another. Since there are only a finite number of possible matchings, we might hope this procedure would eventually lead to a stable matching. However, using the preferences already described in Example 2.4, we can see that this need not be the case, because the procedure can form a cycle, and thus repeat itself indefinitely without reaching a stable matching.

Example 2.4 (continued) (Knuth)
Recall that the matching μ_1 given by

$$\mu_1 = \begin{matrix} w_1 & w_2 & w_3 \\ m_1 & m_2 & m_3 \end{matrix}$$

is unstable, since (m_1, w_2) is a blocking pair.

We can construct a new matching, μ_2, by divorcing the pairs (m_1, w_1) and (m_2, w_2) and marrying m_1 with w_2, as follows:

$$\mu_2 = \begin{matrix} w_1 & w_2 & w_3 \\ m_2 & m_1 & m_3 \end{matrix}.$$

We can see that m_3 prefers w_2 to w_3 and m_3 is the favorite man of w_2. Then μ_2 is also unstable.

Again, we can construct a new matching, μ_3, as follows:

$$\mu_3 = \begin{matrix} w_1 & w_2 & w_3 \\ m_2 & m_3 & m_1 \end{matrix},$$

which is blocked by (m_3, w_1). Making the divorces and the new marriages, we get

$$\mu_4 = \frac{w_1 \quad w_2 \quad w_3}{m_3 \quad m_2 \quad m_1},$$

which is blocked by (m_1, w_1). If we continue with this procedure, we get μ_1 in the next step, instead of a stable matching.

Nevertheless,

$$\mu_5 = \frac{w_1 \quad w_2 \quad w_3}{m_3 \quad m_1 \quad m_2} \quad \text{and} \quad \mu_6 = \frac{w_1 \quad w_2 \quad w_3}{m_1 \quad m_3 \quad m_2}$$

are stable. (Note that a different choice of blocking pairs would converge to a stable matching in this example. Until recently it was an open question whether there exists a rule for choosing blocking pairs that never cycles, that is, that always leads to a stable matching. It turns out that there is: See Theorem 2.33 in Section 2.6.)

Gale and Shapley described the following algorithm, which produces a stable matching starting from any preference lists.

Theorem 2.8 (Gale and Shapley). *A stable matching exists for every marriage market.*

Proof: A procedure for producing a stable matching for any marriage market follows.

To start, each man proposes to his favorite woman, that is, to the first woman on his preference list of acceptable women. Each woman rejects the proposal of any man who is unacceptable to her, and each woman who receives more than one proposal rejects all but her most preferred of these. Any man whose proposal is not rejected at this point is kept "engaged."

At any step any man who was rejected at the previous step proposes to his next choice (i.e., to his most preferred woman among those who have not yet rejected him), so long as there remains an acceptable woman to whom he has not yet proposed. (If at any step of the procedure a man has already proposed to, and been rejected by, all of the women he finds acceptable, then he issues no further proposals.) Each woman receiving proposals rejects any from unacceptable men, and also rejects all but her most preferred among the group consisting of the new proposers together with any man she may have kept engaged from the previous step.

The algorithm stops after any step in which no man is rejected. At this point, every man is either engaged to some woman or has been rejected by every woman on his list of acceptable women. The marriages are now consummated, with each man being matched to the woman to whom he is engaged. Women who did not receive any acceptable proposal, and men who were rejected by all women acceptable to them, will stay single.

This completes the description of the algorithm, except that we have described it as if all agents have strict preferences. The modification required in case some man or woman is indifferent between two or more possible mates is simple. At any step of the algorithm at which some agent must indicate a choice between two mates who are equally well liked, introduce some fixed "tie-breaking" rule (e.g., when an agent is indifferent, proceed as if the preferences are according to alphabetical order of family names, or as if agents prefer mates who are closer to them in age, etc.). Such a tie-breaking rule therefore specifies, arbitrarily, to which woman a man will propose when he is indifferent about his next proposal, and which man a woman will keep engaged when she is indifferent among more than one most favored suitors.

The algorithm must eventually stop because there are only a finite number of men and women, and no man proposes more than once to any woman. The outcome that it produces is a matching, since each man is engaged at any step to at most one woman, and each woman is engaged at any step to at most one man. Furthermore, this matching is individually rational, since no man or woman is ever engaged to an unacceptable partner.

To see that the matching μ produced by the algorithm is stable, suppose some man m and woman w are not married to each other at μ, but m prefers w to his own mate at μ. Then woman w must be acceptable to man m, and so he must have proposed to her before proposing to his current mate (or before being rejected by all of the women he finds acceptable). Since he was not engaged to w when the algorithm stopped, he must have been rejected by her in favor of someone she liked at least as well. Therefore w is matched at μ to a man whom she likes at least as well as man m, since preferences are transitive (and hence acyclic), and so m and w do not block the matching μ. Since the matching is not blocked by any individual or by any pair, it is stable.

We call this algorithm a "deferred acceptance" procedure, to emphasize the fact that women are able to keep the best available man at any step engaged, without accepting him outright. We should emphasize that, for the moment, we present this algorithm only to show that stable matchings always exist. That is, although the algorithm is presented as if at each step the men and women take certain actions, we will not consider until Chapter 4 whether they would be well advised to take those actions, and consequently whether it is reasonable for us to expect that they would act as described, if the rules for making and accepting proposals were as described in the algorithm.

To make sure the algorithm is well understood, let us follow it through an example.

Example 2.9: An example of the deferred acceptance procedure

$$P(m_1) = w_1, w_2, w_3, w_4$$
$$P(m_2) = w_4, w_2, w_3, w_1$$
$$P(m_3) = w_4, w_3, w_1, w_2$$
$$P(m_4) = w_1, w_4, w_3, w_2$$
$$P(m_5) = w_1, w_2, w_4$$

$$P(w_1) = m_2, m_3, m_1, m_4, m_5$$
$$P(w_2) = m_3, m_1, m_2, m_4, m_5$$
$$P(w_3) = m_5, m_4, m_1, m_2, m_3$$
$$P(w_4) = m_1, m_4, m_5, m_2, m_3.$$

First step: m_1, m_4, and m_5 propose to w_1, and m_2 and m_3 propose to w_4; w_1 rejects m_4 and m_5 and keeps m_1 engaged; w_4 rejects m_3 and keeps m_2 engaged. We indicate this in the following manner:

$$\begin{array}{cccc} w_1 & w_2 & w_3 & w_4 \\ m_1 & & & m_2 \end{array}.$$

Second step: m_3, m_4, and m_5 propose to their second choice, that is, to w_3, w_4, and w_2, respectively; w_4 rejects m_2 and keeps m_4 engaged:

$$\begin{array}{cccc} w_1 & w_2 & w_3 & w_4 \\ m_1 & m_5 & m_3 & m_4 \end{array}.$$

Third step: m_2 proposes to his second choice, w_2, who rejects m_5 and keeps m_2 engaged:

$$\begin{array}{cccc} w_1 & w_2 & w_3 & w_4 \\ m_1 & m_2 & m_3 & m_4 \end{array}.$$

Fourth step: m_5 proposes to his third choice, w_4, who rejects m_5 and continues with m_4 engaged. Since m_5 has been rejected by every woman on his list of acceptable women, he stays single, that is, matched with himself, and the stable matching obtained is:

$$\mu_M = \begin{array}{ccccc} w_1 & w_2 & w_3 & w_4 & (m_5) \\ m_1 & m_2 & m_3 & m_4 & m_5 \end{array}.$$

We have called this matching μ_M to draw attention to the fact that it results from the procedure in which proposals are made by the men. Since men and women play precisely symmetrical roles in the marriage market, and since they play different roles in the procedure just described, we could have described another algorithm in which the roles of men and women were reversed. This would result in a matching μ_W that would also be stable. These two stable matchings will not typically be the same. For the marriage market of Example 2.9, the stable matching obtained when the women propose to the men is

$$\mu_W = \frac{w_4 \ w_1 \ w_2 \ w_3 \ (m_5)}{m_1 \ m_2 \ m_3 \ m_4 \ m_5}.$$

Note in this example that *all* the men like μ_M at least as well as μ_W, and *all* the women prefer μ_W to μ_M. At first glance, it might appear natural that a procedure that treats the two sides of the market in different ways should give rise to matchings that systematically favor one side of the market. But on second thought, it isn't clear whether this has any meaning. After all, it is fairly clear that at least for some configurations of preferences, the men are competing with other men, and the women are competing with other women for desirable spouses. And although there is this conflict of interest between agents on the same side of the market (who all might like to be matched, say, to the same mate), there is a lot of common interest between agents on opposite sides of the market, who are after all interested in being matched to one another. So when we look at two different matchings, we should expect to find that one of them is preferred by some men and some women, as is the other. This is the sense in which it isn't obvious that the idea of systematically favoring one *side* of the market has any meaning. The observation in Example 2.9 that men prefer μ_M and women prefer μ_W might just be an accident.

It turns out not to be an accident, but an example of a quite general phenomenon. This is one of the most surprising and important discoveries about the class of two-sided markets of which the marriage market is a member. When all agents have strict preferences, there *are* systematic elements of common interests among the men (and among the women), and systematic conflicts of interest between men and women, even in cases in which all men are competing for the same woman, and all women are competing for the same man. This is a sufficiently important result that we will discuss it from a number of angles.

Let us begin with an example in which all the men prefer the same woman and all the women prefer the same man, to see what happens in such a case.

*Example 2.10: An example in which all men prefer the
same woman, and all women prefer the same man*

$$P(m_1) = w_1, w_2, w_3 \qquad P(w_1) = m_1, m_2, m_3$$
$$P(m_2) = w_1, w_2, w_3 \qquad P(w_2) = m_1, m_3, m_2$$
$$P(m_3) = w_1, w_3, w_2 \qquad P(w_3) = m_1, m_2, m_3$$

In this example, woman w_1 is the first choice of every man, and man m_1 is the first choice of every woman. So the sense in which men are competing with men, and women are competing with women is very real. No two men would agree on what is the best matching, since each man's favorite matching is one at which he is married to woman w_1. (Recall that an agent who has strict preferences is indifferent between two individually rational matchings if and only if he has the same mate at both of them.) Similarly, no two women agree on what is the best matching. But let us now turn our attention to the set of *stable* matchings. Any matching that does not pair m_1 with w_1 is unstable, since m_1 and w_1 are each other's first choice, and so form a blocking pair for any matching at which they are not mates. Consequently there are only two stable matchings. These are

$$\mu_M = \begin{matrix} w_1 & w_2 & w_3 \\ m_1 & m_2 & m_3 \end{matrix} \quad \text{and} \quad \mu_W = \begin{matrix} w_1 & w_3 & w_2 \\ m_1 & m_2 & m_3 \end{matrix}.$$

So when we confine our attention to the stable matchings, the disagreement among the men disappears. All the men like the matching μ_M that results from the algorithm when the men propose at least as well as the matching μ_W that results when the women propose. Man m_1 is indifferent between the two matchings, and the other men prefer μ_M. Similarly, the women are not in any disagreement that μ_W is the optimal stable matching.

Let's start to examine this phenomenon more formally. Let $\mu >_M \mu'$ denote that all men like μ at least as well as μ', with at least one man preferring μ to μ' outright, that is, that $\mu(m) \geq_m \mu'(m)$ for all m, and $\mu(m) >_m \mu'(m)$ for at least one man m. Let $\mu \geq_M \mu'$ denote that either $\mu >_M \mu'$ or that all men are indifferent between μ and μ'. Note that the relation $\geq_M$, which represents the *common* preferences of the men, is unlike an individual's preference ordering. It is *partial*, rather than complete, in that not all matchings can be compared. (Specifically, matchings μ and μ' cannot be compared by $\geq_M$ if some men prefer μ and others prefer μ'.) However, like individual preferences, $\geq_M$ is transitive, since if all men like μ at least as well as μ' and μ' at least as well as μ'', then all men like μ at least as well as μ''. In a precisely similar way, we define $\geq_W$ and $>_W$ to represent the common preferences of the women over alternative matchings.

Example 2.10 shows that even when there is a lot of disagreement among the men about which is the best matching, there may still be a good deal of agreement about which is the best stable matching. For a given marriage market, we say that a stable matching μ is an *M-optimal stable matching* if every man likes it at least as well as any other stable matching. We can define a W-optimal stable matching similarly. Formally, we have:

Definition 2.11. *For a given marriage market* (M, W, P), *a stable matching* μ *is M-optimal if every man likes it at least as well as any other stable matching; that is, if for every other stable matching* μ', $\mu \geq_M \mu'$. *Similarly, a stable matching* ν *is W-optimal if every woman likes it at least as well as any other stable matching, that is, if for every other stable matching* ν', $\nu \geq_W \nu'$.

Each individual agent compares alternative matchings in terms of his or her preferences for his or her own mates at those matchings. So in examining the set of stable matchings, an agent is involved in comparing those mates whom he or she might have at some stable matching. Define a woman w and a man m to be *achievable* for each other in a marriage market (M, W, P) if m and w are paired at some stable matching. In a marriage market in which all men and women have strict preferences, each man and woman who has any achievable mates has (only) one favorite among these. Consequently, in such a market an M-optimal stable matching must match each man to his most preferred achievable woman, and a W-optimal stable matching must match each woman to her most preferred achievable man. So, when preferences are strict, there can be at most one M-optimal stable matching and one W-optimal stable matching. The following theorem of Gale and Shapley shows that such optimal stable matchings do in fact exist.

Theorem 2.12 (Gale and Shapley). *When all men and women have strict preferences, there always exists an M-optimal stable matching, and a W-optimal stable matching. Furthermore, the matching* μ_M *produced by the deferred acceptance algorithm with men proposing is the M-optimal stable matching. The W-optimal stable matching is the matching* μ_W *produced by the algorithm when the women propose.*

Proof: When all men and women have strict preferences, we will show that in the deferred acceptance algorithm with men proposing, no man is ever rejected by an achievable woman. Consequently the stable matching μ_M that is produced matches each man to his most preferred achievable woman, and is therefore the (unique) M-optimal stable matching.

The proof is by induction. Assume that up to a given step in the procedure no man has yet been rejected by a woman who is achievable for him. At this step, suppose woman w rejects man m. If she rejects m as unacceptable, then she is unachievable for him, and we are done. If she rejects m in favor of man m', whom she keeps engaged, then she prefers m' to m. We must show that w is not achievable for m.

We know m' prefers w to any woman except for those who have previously rejected him, and hence (by the inductive assumption) are unachievable for him. Consider a hypothetical matching μ that matches m to w and everyone else to an achievable mate. Then m' prefers w to his mate at μ. So the matching μ is unstable, since it is blocked by m' and w, who each prefer the other to their mate at μ. Therefore there is no stable matching that matches m and w, and so they are unachievable for each other, which completes the proof.

Thus, when preferences are strict, the agents on one side of the market have a common interest regarding the set of stable matchings, since they are in agreement on the best stable matching. It turns out that agents on opposite sides of the market have opposite interests in this regard, and the optimal stable matching for one side of the market is the worst stable matching for agents on the other side of the market. This opposition of interests between the two sides of the market can be observed not only in comparing the optimal stable matchings for each side of the market, but in comparing any two stable matchings. Any stable matching that is better for all the men is worse for all the women, and vice versa. We can state this formally as follows.

Theorem 2.13 (Knuth).   *When all agents have strict preferences, the common preferences of the two sides of the market are opposed on the set of stable matchings: if μ and μ' are stable matchings, then all men like μ at least as well as μ' if and only if all women like μ' at least as well as μ. That is, $\mu >_M \mu'$ if and only if $\mu' >_W \mu$.*

An immediate consequence of this theorem is the following.

Corollary 2.14.   *When all agents have strict preferences, the M-optimal stable matching is the worst stable matching for the women; that is, it matches each woman with her least preferred achievable mate. Similarly, the W-optimal stable matching matches each man with his least preferred achievable mate.*

Proof of the theorem: Let μ and μ' be stable matchings such that $\mu >_M \mu'$. We will show that $\mu' >_W \mu$.

Suppose it is not true that $\mu' >_W \mu$. Then there must be some woman w who prefers μ to μ'. (This follows from the strict preferences of the women, together with the fact that at least one woman has a different mate at μ and μ' since at least one man does). Then woman w has a different mate at μ and μ', and consequently so does man $m = \mu(w)$. (Since all stable matchings must be individually rational, the fact that w prefers $\mu(w)$ to $\mu'(w)$ implies w is not single at μ.) Since man m also has strict preferences, m and w form a blocking pair for the matching μ'. This contradicts the assumption that μ' is stable. Therefore $\mu' >_W \mu$, as required.

In Theorems 2.12 and 2.13 we have been careful to specify that all agents must have strict preferences. If some agents may be indifferent between possible mates, these results need not hold. The following example is one in which no M-optimal and W-optimal stable matchings exist.

Example 2.15: An example in which not all agents have strict preferences

$$P(m_1) = [w_2, w_3], w_1 \qquad P(w_1) = m_1, m_2, m_3$$
$$P(m_2) = w_2, w_1 \qquad P(w_2) = m_1, m_2$$
$$P(m_3) = w_3, w_1 \qquad P(w_3) = m_1, m_3$$

The stable matchings are

$$\mu_1 = \begin{matrix} w_1 & w_2 & w_3 \\ m_2 & m_1 & m_3 \end{matrix} \quad \text{and} \quad \mu_2 = \begin{matrix} w_1 & w_2 & w_3 \\ m_3 & m_2 & m_1 \end{matrix},$$

but there are no optimal stable matchings since $\mu_1(m_3) >_{m_3} \mu_2(m_3)$ but $\mu_2(m_2) >_{m_2} \mu_1(m_2)$, and $\mu_1(w_2) >_{w_2} \mu_2(w_2)$ but $\mu_2(w_3) >_{w_3} \mu_1(w_3)$.

So the pattern we are starting to see in the set of stable matchings – of common interests among agents on the same side of the market and conflicting interests between the two sides of the market – is related to agents having strict preferences. That is, the pattern is related to their ability to distinguish between alternative (acceptable) mates.

Of course there are many reasons, in practical situations of the sort modeled here, to expect that agents might not have the ability to distinguish among all alternatives facing them. Perhaps the most important such reason is that agents might have little information about some alternatives, and consequently be indifferent between them. (Note in this connection that our definition of strict preferences does not rule out indifference between unacceptable mates.) However in situations where the agents have a great deal of information, we might even consider the case in which they have strict preferences to be typical. Loosely speaking, the reason is that indifference is in some sense a "knife edge" phenomenon; if an agent is indifferent between two alternatives, a small improvement in one of them would presumably cause him or her to prefer it to the other. But if an agent clearly prefers one alternative to another, then a sufficiently small improvement in the less preferred alternative will presumably leave the preferences unchanged. In this sense, an example with strict preferences is more robust, less special, than an example with indifference. A related point of view is that when an agent in an economic model is modeled as being indifferent between two alternatives, this is as likely to reflect a shortage of information on the part of the modeler as it is to reflect something about the agent's behavior – if more were known about the agent's tastes, we might be able to model his or her choice behavior more accurately.

In any event, the phenomena displayed in Theorems 2.12 and 2.13 are sufficiently striking that we would like to better understand the underlying causes of the conflict and coincidence of common interests we have noted with regard to the M- and W-optimal stable matchings.

Let us begin by informally reconsidering what we have learned so far. Suppose all men and women have strict preferences, and that we approach the men in a particular marriage market and say to each of them, "Point to your most preferred mate." Then it might be that more than one man would point to a given woman (as in Example 2.10). But suppose we instead instruct the men by saying to them, "Be realistic, and point to your most preferred mate among those who might actually agree to marry *you*, given the competition from the other men. Specifically, point to your most preferred mate among those whom you could be matched with at a stable matching, that is, your most preferred achievable mate." Then Theorem 2.12 tells us two surprising things. First, *no two men point to the same woman,* so there is a matching at which each man is married to the woman he pointed to. Second, *this matching is stable.* It is, of course, the M-optimal stable matching, and the W-optimal stable matching would be obtained by asking the women to do the pointing. Theorem 2.13 tells us further that we could also obtain the W-optimal stable matching, say, by asking the men to point to their *least* preferred achievable mate.

These same "pointing" phenomena occur even over more restricted choices. We continue to suppose that all agents have strict preferences. For any two matchings μ and μ', ask each man to point to his preferred mate from the two; that is, ask each man m to point to whichever of $\mu(m)$ or $\mu'(m)$ he prefers. For arbitrary matchings μ and μ' it might be that some woman will be pointed to by more than one man, but if μ and μ' are stable, we'll show that this won't occur. So there will be another matching, λ, at which each man is matched to his choice from μ and μ', and this matching will also turn out to be stable. Every man likes λ at least as well as either μ or μ', and consequently (by Theorem 2.13) every woman likes both μ and μ' at least as well as λ. In the same way, there will be a stable matching ν that is worse for the men and better for the women.

More formally, when preferences are strict we can define, for any two matchings μ and μ', the following function on the set $M \cup W$. Let $\lambda = \mu \vee_M \mu'$, be defined by $\lambda(m) = \mu(m)$ if $\mu(m) >_m \mu'(m)$ and $\lambda(m) = \mu'(m)$ otherwise, for all m in M; $\lambda(w) = \mu(w)$ if $\mu(w) <_w \mu'(w)$ and $\lambda(w) = \mu'(w)$ otherwise, for all w in W. This is the "pointing function" described in the previous paragraphs – it assigns each man his more preferred mate from μ and μ', and it assigns each woman her less preferred mate. In a precisely similar way we can define the function $\nu = \mu \wedge_M \mu'$, which gives each man his less preferred mate and each woman her more preferred mate.

There are two ways in which these functions λ and ν might fail to be matchings. Consider λ. First, it might be that $\lambda(m) = \lambda(m')$ for two different men m and m'; that is, λ might assign the same woman to two different men. Second, there might be a man m and woman w such that $\lambda(m) = w$, but $\lambda(w) \neq m$; that is, it might be that giving each man the more preferred of his mates at μ and μ' is not identical to giving each woman the less preferred of her mates. And, of course, even when λ and ν are matchings, they might not be stable. However when μ and μ' are stable, it turns out that λ and ν will always be stable matchings. (The fact that the operations $\vee_M$ and $\wedge_M$ each produce a stable matching from a pair of stable matchings implies that the set of stable matchings is an example of an algebraic structure called a lattice. This is discussed at greater length in Section 3.1.1.)

Theorem 2.16: Lattice theorem (Conway). *When all preferences are strict, if μ and μ' are stable matchings, then the functions $\lambda = \mu \vee_M \mu'$ and $\nu = \mu \wedge_M \mu'$ are both matchings. Furthermore, they are both stable.*

Proof: First we show λ is a matching, by showing that $\lambda(m) = w$ if and only if $\lambda(w) = m$. The stability of μ and μ' guarantees the "only if" direction, that is, if $\lambda(m) = w$ then $\lambda(w) = m$. To see the "if" direction, let $M' = \{m$ such that $\lambda(m)$ is in $W\} = \{m$ such that μ or $\mu'(m)$ is in $W\}$. By the only if direction $\lambda(M')$ is contained in $\{w$ such that $\lambda(w)$ is in $M\}$ which (by definition of λ) equals $W' \equiv \{w$ such that $\mu(w)$ and $\mu'(w)$ are in $M\}$, which is the same size as $\mu(W')$. But $\lambda(M')$ is the same size as M' (since $\lambda(m) = \lambda(m') = w$ only if $m = m' = \lambda(w)$), which is at least as large as $\mu(W')$, so $\lambda(M')$ and W' are the same size and $\lambda(M') = W'$. Hence for w in W', $\lambda(w) = m$ for some m in M', so $\lambda(w) = m$. If w is not in W' then $\lambda(w) = w$. So if $\lambda(w) = m$ then $\lambda(m) = w$. That ν is a matching follows from a symmetric argument.

To see that λ is stable, suppose (m, w) blocks λ. Then $w >_m \lambda(m)$, from which it follows that $w >_m \mu(m)$ and $w >_m \mu'(m)$. On the other hand, $m >_w \lambda(w)$. Hence, (m, w) blocks μ if $\lambda(w) = \mu(w)$, or μ' if $\lambda(w) = \mu'(w)$. In either case we get a contradiction, since μ and μ' are stable. By the symmetric argument ν is stable also.

Note that the existence of M- and W-optimal stable matchings can be deduced from the lattice theorem. To see this, consider a matching μ that is not M-optimal. Then we can find another stable matching μ' that is preferred by at least one man, and all men agree that the stable matching $\lambda = \mu \vee_M \mu'$ is at least as good as either of the two original matchings. If λ is M-optimal we are done, and otherwise we can proceed in the same way, increasing the welfare of at least some men at each step. Since there are only a finite number of stable matchings, this process ends when λ is M-optimal.

Example 2.15 illustrated a situation in which no optimal stable matchings exist when preferences are not strict, and hence the set of stable matchings is not a lattice. Example 2.17 illustrates the lattice property of the set of stable matchings.

Example 2.17: The lattice of stable matchings (Knuth)

$$P(m_1) = w_1, w_2, w_3, w_4 \qquad P(w_1) = m_4, m_3, m_2, m_1$$
$$P(m_2) = w_2, w_1, w_4, w_3 \qquad P(w_2) = m_3, m_4, m_1, m_2$$
$$P(m_3) = w_3, w_4, w_1, w_2 \qquad P(w_3) = m_2, m_1, m_4, m_3$$
$$P(m_4) = w_4, w_3, w_2, w_1 \qquad P(w_4) = m_1, m_2, m_3, m_4$$

There are ten stable matchings where w_1, w_2, w_3, and w_4 are matched respectively to

$$m_1 \ m_2 \ m_3 \ m_4 \tag{1}$$
$$m_2 \ m_1 \ m_3 \ m_4 \tag{2}$$
$$m_1 \ m_2 \ m_4 \ m_3 \tag{3}$$
$$m_2 \ m_1 \ m_4 \ m_3 \tag{4}$$
$$m_3 \ m_1 \ m_4 \ m_2 \tag{5}$$
$$m_2 \ m_4 \ m_1 \ m_3 \tag{6}$$
$$m_3 \ m_4 \ m_1 \ m_2 \tag{7}$$
$$m_4 \ m_3 \ m_1 \ m_2 \tag{8}$$
$$m_3 \ m_4 \ m_2 \ m_1 \tag{9}$$
$$m_4 \ m_3 \ m_2 \ m_1. \tag{10}$$

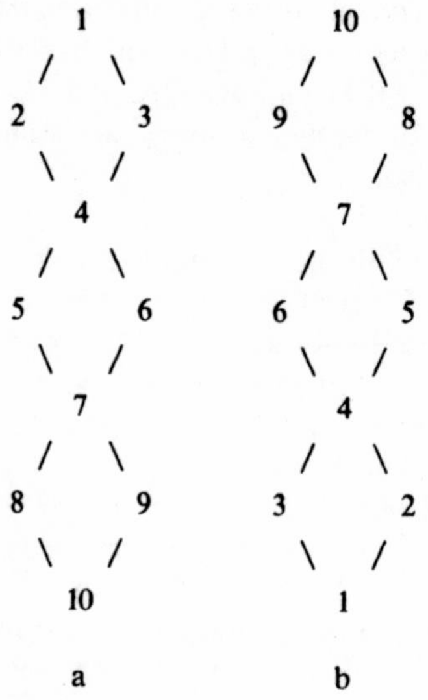

Figure 2.1.

We can see that

$$\mu_2 \underset{M}{\wedge} \mu_3 = \mu_4 \qquad \mu_2 \underset{M}{\vee} \mu_3 = \mu_1$$

$$\mu_5 \underset{M}{\wedge} \mu_6 = \mu_7 \qquad \mu_5 \underset{M}{\vee} \mu_6 = \mu_4$$

$$\mu_8 \underset{M}{\wedge} \mu_9 = \mu_{10} \qquad \mu_8 \underset{M}{\vee} \mu_9 = \mu_7$$

where μ_j is the matching described in (j), for all $j = 1, \ldots, 10$. (Note that for any two stable matchings μ and μ' such that $\mu >_M \mu'$, $\mu = \mu \vee_M \mu'$ and $\mu' = \mu \wedge_M \mu'$.) The M-optimal stable matching is μ_1 and the W-optimal stable matching is μ_{10}.

We can graphically represent the set of stable matchings by one of the lattices shown in Figure 2.1, where the numbers in the vertices indicate the matchings. A matching μ is higher up in the lattice of 2.1a (respectively Figure 2.1b) than another matching μ' if and only if $\mu >_M \mu'$ (respectively, if and only if $\mu' >_W \mu$).

The figure illustrates what the lattice theorem has to say about the extent to which the common interests of the men coincide on the set of stable matchings, and conflict with the common interests of the women. This coincidence and conflict of common interests is partial rather than total. There are stable matchings, such as μ_5 and μ_6 in the example, such that some men and women prefer μ_5, while other men and women prefer μ_6. But all men agree that μ_4 is at least as good as either μ_5 or μ_6, which are both at least as good as μ_7. And all women have the reverse preferences over μ_4 and μ_7.

2.4　Extending the domain of the theory: some examples

So far we have been concerned with the set of stable outcomes within a particular marriage market. In this section we begin to consider, via some examples, some comparisons between different markets. The theory underlying these examples is developed in Section 2.5.

One kind of comparison concerns the effect on the men and women if one of the men, say, should change his preferences by *extending* his list of acceptable women to include some woman whom he previously found unacceptable. We will see that such a change cannot harm the women and cannot help the (other) men. A second kind of comparison concerns the effect of introducing some additional people into a marriage market. We will see that if we add a new woman, for example, this cannot harm the men and cannot help the (other) women.

Example 2.18: The effect of extending the mens' preferences
Consider the market with men $M = \{m_1, \ldots, m_6\}$, women $W = \{w_1, \ldots, w_5\}$ and preferences P given by:

$$P(m_1) = w_1, w_3 \qquad P(w_1) = m_2, m_1, m_6$$
$$P(m_2) = w_2, w_4 \qquad P(w_2) = m_6, m_1, m_2$$
$$P(m_3) = w_4, w_3 \qquad P(w_3) = m_3, m_4, m_1, m_5$$
$$P(m_4) = w_3, w_4 \qquad P(w_4) = m_4, m_3, m_2$$
$$P(m_5) = w_5 \qquad P(w_5) = m_5$$
$$P(m_6) = w_1, w_4.$$

The M- and W-optimal stable matchings are given by:

$$\mu_M = \begin{matrix} w_1 & w_2 & w_3 & w_4 & w_5 & (m_6) \\ m_1 & m_2 & m_4 & m_3 & m_5 & m_6 \end{matrix}, \qquad \mu_W = \begin{matrix} w_1 & w_2 & w_3 & w_4 & w_5 & (m_6) \\ m_1 & m_2 & m_3 & m_4 & m_5 & m_6 \end{matrix}.$$

Suppose now that some of the men decide to extend their lists of acceptable women yielding the new preference profile P':

$$P'(m_1) = w_1, w_3, w_2$$
$$P'(m_2) = w_2, w_4, w_1$$
$$P'(m_3) = w_4, w_3, w_2$$

$$P'(m_4) = w_3, w_4$$
$$P'(m_5) = w_5, w_3$$
$$P'(m_6) = w_1, w_4, w_2.$$

In this case the M- and W-optimal stable matchings are:

$$\mu'_M = \begin{array}{cccccc} w_1 & w_2 & w_3 & w_4 & w_5 & (m_1) \\ m_2 & m_6 & m_4 & m_3 & m_5 & m_1 \end{array},\qquad \mu'_W = \begin{array}{cccccc} w_1 & w_2 & w_3 & w_4 & w_5 & (m_1) \\ m_2 & m_6 & m_3 & m_4 & m_5 & m_1 \end{array}.$$

Under the original preferences P, no man is worse off and no woman is better off at μ_M than at μ'_M.

However men m_3 and m_4 are better off under μ'_M than under μ_W and the women w_3 and w_4 are better off under μ_W than under μ'_M. Both μ_W and μ'_M map $\{m_3, m_4\}$ onto $\{w_3, w_4\}$. Note also that in each of the markets (M, W, P) and (M, W, P'), everyone is either matched at both the M and W-stable matchings or unmatched at both.

Example 2.19: The effect of adding another woman
Consider the market with $M = \{m_1, m_2, m_3\}$, $W = \{w_1, w_2, w_3\}$, and preferences P given by:

$$P(m_1) = w_1, w_3 \qquad P(w_1) = m_1, m_3$$
$$P(m_2) = w_3, w_2 \qquad P(w_2) = m_2$$
$$P(m_3) = w_1, w_3 \qquad P(w_3) = m_3, m_2.$$

There is a single stable matching in this example:

$$\mu_M = \mu_W = \begin{array}{ccc} w_1 & w_2 & w_3 \\ m_1 & m_2 & m_3 \end{array}.$$

Suppose woman w_4 now enters the market and m_1 prefers her to w_1. The new market is given by $M' = M$, $W' = \{w_1, w_2, w_3, w_4\}$, and P' given by:

$$P'(m_1) = w_4, w_1, w_3 \qquad P'(w_1) = m_1, m_3$$
$$P'(m_2) = w_3, w_2 \qquad P'(w_2) = m_2$$
$$P'(m_3) = w_1, w_3 \qquad P'(w_3) = m_3, m_2$$
$$\qquad\qquad\qquad\qquad P'(w_4) = m_2, m_1.$$

Again there is a single stable matching under P':

$$\mu'_M = \mu'_W = \begin{array}{cccc} w_1 & w_2 & w_3 & w_4 \\ m_3 & (w_2) & m_2 & m_1 \end{array}.$$

Under the preferences P', all the men are better off at μ'_M than at μ_M.

In the following section, we will discover the general rules that govern the various observations we have made about these examples. We will concentrate on the simplest case, in which all agents have strict preferences.

2.5 A concise mathematical treatment when preferences are strict

In developing the theory to deal with the kind of observations made about Examples 2.18 and 2.19, we will prove a powerful lemma that will enable us to make some new observations, and also allow us to reprove almost all of our previously obtained results, using only the properties of stability and optimality. Therefore this section gives a unified treatment of the theory developed so far, in a way that casts previous results in a new light. For simplicity, we will assume throughout this section that all preferences are strict.

We will need some notation to discuss what happens when a man or woman extends his or her list of acceptable people by adding people to the end of the original list of acceptable people, as in Example 2.18. We will write $P'_m \geq P_m$ if P'_m is such an extension of P_m. Similarly, we will write $P'_w \geq P_w$ and finally we will write $P' \geq_M P$ if $P'_m \geq P_m$ for all m in M.

Lemma 2.20: Decomposition lemma (Gale and Sotomayor). *Let μ and μ' be, respectively, stable matchings in markets (M, W, P) and (M, W, P') with $P' \geq_M P$, and all preferences are strict. Let $M(\mu')$ be the set of men who prefer μ' to μ under P and let $W(\mu)$ be the set of women who prefer μ to μ'. Then μ' and μ map $M(\mu')$ onto $W(\mu)$. (That is, both μ' and μ match any man who prefers μ' to a woman who prefers μ, and vice versa.)*

Proof: Suppose $m \in M(\mu')$. Then $\mu'(m) >_m \mu(m) \geq_m m$, under P so $\mu'(m) \in W$. Setting $w = \mu'(m)$, we cannot have $\mu'(w) >_w \mu(w)$ for then (m, w) would block μ. Hence, since the preferences are strict, $w \in W(\mu)$ and so $\mu'(M(\mu'))$ is contained in $W(\mu)$.

On the other hand, if $w \in W(\mu)$ then $\mu(w) >_w \mu'(w) \geq_w w$, so $\mu(w) \in M$. Letting $\mu(w) = m$, we see that we cannot have $\mu(m) >_m \mu'(m)$ under P' or (m, w) would block μ'. Hence, since the preferences are strict, $\mu'(m) >_m \mu(m) = w >_m m$ under P' and P. So $m \in M(\mu')$ and $\mu(W(\mu))$ is contained in $M(\mu')$.

Since μ and μ' are one-to-one and $M(\mu')$ and $W(\mu)$ are finite, the conclusion follows.

Note that when we consider two stable matchings from the *same* marriage market (so that $P = P'$) the decomposition lemma implies for example

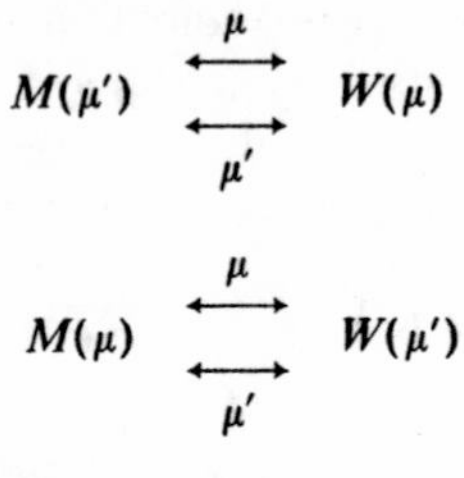

$$M(\mu') \quad \overset{\mu}{\underset{\mu'}{\longleftrightarrow}} \quad W(\mu)$$

$$M(\mu) \quad \overset{\mu}{\underset{\mu'}{\longleftrightarrow}} \quad W(\mu')$$

Figure 2.2.

that if $\mu(m) = w$ and $\mu'(m) = w'$ and m prefers μ to μ', then both w and w' will prefer μ' to μ. That is, both μ and μ' decompose the men and women as illustrated in Figure 2.2.

We state this result formally below, because it will be very useful in what follows.

Corollary 2.21: Decomposition lemma when $P = P'$ (Knuth).   *Let μ and μ' be stable matchings in (M, W, P), where all preferences are strict. Let $M(\mu)$ be the set of men who prefer μ to μ' and $W(\mu)$ be the set of women who prefer μ to μ'. Analogously define $M(\mu')$ and $W(\mu')$. Then μ and μ' map $M(\mu')$ onto $W(\mu)$ and $M(\mu)$ onto $W(\mu')$.*

This result allows us to give the following alternate proof of Theorem 2.13.

Alternate proof of Theorem 2.13: $\mu' >_M \mu$ under P if and only if $M(\mu)$ is empty and $M(\mu')$ is nonempty. This is equivalent to $\mu(m) = \mu'(m)$ for all m in $M - M(\mu')$ and $W(\mu)$ is nonempty, which in turn is satisfied if and only if $\mu(w) = \mu'(w)$ for all w in $W - W(\mu)$ and $\mu(w) >_w \mu'(w)$ for some w. But this is equivalent to $\mu >_W \mu'$.

The next theorem concerns a set of people who are indifferent between all stable matchings. The result does not seem obvious although, as we will see, its proof is a simple consequence of the decomposition lemma when $P = P'$.

Theorem 2.22.   *In a market (M, W, P) with strict preferences, the set of people who are single is the same for all stable matchings.*

Proof: Suppose m was matched under μ' but not under μ. Then $m \in M(\mu')$, but from the decomposition lemma with $P = P'$, μ maps $W(\mu)$

onto $M(\mu')$, so m is also matched under μ, which gives the necessary contradiction.

Lemma 2.23. *With the assumptions and notation of Lemma 2.20 and Theorem 2.16, we have*

$$\lambda = \mu \underset{M}{\vee} \mu', \text{ under } P, \text{ is a matching and is stable for } (M, W, P)$$

$$\nu = \mu \underset{W}{\vee} \mu', \text{ under } P, \text{ is a matching and is stable for } (M, W, P').$$

Proof: By definition, $\mu \vee_M \mu'$ must agree with μ' on $M(\mu')$ and $W(\mu)$ and with μ otherwise. By the decomposition lemma, λ is therefore a matching. Further, for $m \in M(\mu')$, $\mu'(m) >_m \mu(m) \geq_m m$ under P so $\mu'(m)$ is acceptable to m under P, and λ is individually rational in (M, W, P). Suppose now that some (m, w) blocks λ and that $m \in M(\mu')$ so $w >_m \mu'(m)$. Then certainly $w >_m \mu(m)$ so if $w \in W(\mu)$, then $m >_w \mu'(w)$ and μ' would be blocked, and if $w \in W - W(\mu)$, then $m >_w \mu(w)$ and μ would be blocked. On the other hand, if $m \in M - M(\mu')$, then $\mu(m) \geq_m \mu'(m)$ so (m, w) would block μ or μ' according to whether w is in $W - W(\mu)$ or $W(\mu)$. This proves that λ is a stable matching.

Similarly, ν agrees with μ on $W(\mu) \cup M(\mu')$ and with μ' otherwise. Again, ν is an individually rational matching for (M, W, P') from the decomposition lemma. The stability argument is as in the previous paragraph.

The lattice theorem (Theorem 2.16) is a consequence of Lemma 2.23 when $P = P'$. As noted earlier, the lattice theorem, together with the nonemptiness of the set of stable matchings (Theorem 2.8), in turn implies the existence of the M-optimal and W-optimal stable matchings (Theorem 2.12).

The following theorem concerns the principal question that motivated Example 2.18: What happens to the M-optimal stable matching, for example, if a man extends his list of acceptable spouses?

Theorem 2.24 (Gale and Sotomayor). *Suppose $P' \geq_M P$ and let μ'_M, μ_M, μ'_W, and μ_W be the corresponding optimal matchings. Then under the preferences P the men are not worse off and the women are not better off in (M, W, P) than in (M, W, P'), no matter which of the two optimal matchings are considered. That is,*

$$\mu_M \underset{M}{\geq} \mu'_M \text{ under } P \text{ (so } \mu'_M \underset{W}{\geq} \mu_M \text{ by the stability of } \mu'_M), \quad \text{and}$$

$$\mu'_W \underset{W}{\geq} \mu_W \text{ (so } \mu_W \underset{M}{\geq} \mu'_W \text{ under } P, \text{ by the stability of } \mu_W).$$

Proof: By Lemma 2.23, $\mu_M \vee_M \mu'_M$ under P is stable for (M, W, P), so $\mu_M \geq_M (\mu_M \vee_M \mu'_M) \geq_M \mu'_M$. Also by Lemma 2.23, $\mu_W \vee_W \mu'_W$ under P is stable for (M, W, P'), so $\mu'_W \geq_W (\mu_W \vee_W \mu'_W) \geq_W \mu_W$.

The next theorem addresses the issue raised in Example 2.19 of what happens if additional women are added to the market for example. Note that the decomposition lemma does not apply, since the preferences of the men in a market containing new women are not extensions of their original preferences (since the new women may be preferred by some men to other acceptable women). However the proof makes use of the results of Theorem 2.24, via the observation that adding new women to the market is in a certain sense equivalent to extending the preferences of the *women*. This is because we can regard the original market (in which the new women are not present) as equivalent to a market in which they are present, but are unavailable because they find all men unacceptable. The market in which the new women are present is now equivalent to one in which previously unavailable women extend their preferences.

Theorem 2.25 (Gale and Sotomayor). *Suppose W is contained in W' and μ_M and μ_W are the man and woman optimal matchings, respectively, for (M, W, P). Let μ'_M and μ'_W be the man and woman optimal matchings, respectively, for (M, W', P'), where P' agrees with P on M and W. Then*

$$\mu_W \underset{W}{\geq} \mu'_W \text{ under } P \text{ and } \mu'_W \underset{M}{\geq} \mu_W \text{ under } P', \quad and$$

$$\mu'_M \underset{M}{\geq} \mu_M \text{ under } P' \text{ and } \mu_M \underset{W}{\geq} \mu'_M \text{ under } P.$$

Proof: Denote by P'' the set of preference lists such that P'' agrees with P' on $M \cup W$ and $P''(w) = w$ for all $w \in W' - W$. Let μ''_M and μ''_W be the optimal matchings under P''. Since no man is acceptable to any women in $W' - W$ under P'', μ''_W agrees with μ_W on $M \cup W$ and μ''_M agrees with μ_M on $M \cup W$. Now observe that $P' \geq_W P''$. So we can apply Theorem 2.24 and obtain that $\mu''_W \geq_W \mu'_W$ under P'', so $\mu_W \geq_W \mu'_W$. Similarly, $\mu'_W \geq_M \mu''_W$ so $\mu'_W \geq_M \mu_W$ under P', and $\mu'_M \geq_M \mu_M$ under P'. Finally, $\mu''_M \geq_{W'} \mu'_M$ under P'', so $\mu_M \geq_W \mu'_M$ under P.

The previous theorem states that when new women enter the market, no man is hurt at the M-optimal matching. The next theorem says that (unless the new women remain unmatched) there exist some men who are in fact helped in quite a clear way: They are better off at *every* stable matching in the new market than they were at *any* stable matching of the old market. Furthermore (unless these men were all previously unmatched), there are some women who are similarly harmed by the entry of new women into the market.

Theorem 2.26. *Suppose a woman w_0 is added to the market and let μ'_W be the women-optimal stable matching for $(M, W' = W \cup \{w_0\}; P')$, where P' agrees with P on W. Let μ_M be the M-optimal stable matching for (M, W, P). If w_0 is not single under μ'_W, then there exists a nonempty subset of men, S, such that if a man is in S he is better off, and if a woman is in $\mu_M(S)$ she is worse off under any stable matching for the new market than at any stable matching for the original market, under the new (strict) preferences P'.*

Proof: Let $m_0 = \mu'_W(w_0)$. If m_0 is single under μ_M then we can take $S = \{m_0\}$. So suppose $\mu_M(m_0) = w_1 \in W$. It is enough to show that there exists a set of men, S, such that

$$\mu'_W(m) \underset{m}{>} \mu_M(m) \text{ under } P', \text{ for all } m \in S, \quad \text{and}$$

$$\mu'_W(w) \underset{w}{<} \mu_M(w), \text{ for any woman } w \text{ in } \mu_M(S).$$

Construct a directed graph whose vertices are $M \cup W$. There are two types of arcs. If $m \in M$ and $\mu_M(m) = w \in W$, there is an arc from m to w; if $w \in W$ and $\mu'_W(w) = m \in M$, there is an arc from w to m. Let $\bar{M} \cup \bar{W}$ be all vertices that can be reached by a directed path starting from m_0.

Case 1: $\bar{W}$ has a woman $w_{k+1} \in W$ who is single under μ'_W. Let $(m_0, w_1, m_1, w_2, \ldots, w_k, m_k, w_{k+1})$ be the path from m_0 to w_{k+1}. Hence $\mu'_W(m_i) = w_i$ and $\mu_M(m_i) = w_{i+1}$, for all $i = 1, \ldots, k$. We claim that the set $\{m_0, m_1, \ldots, m_k\}$ has the desired properties. In fact, if $\mu_M(w_i) >_{w_i} \mu'_W(w_i)$ for some $i \in \{1, \ldots, k+1\}$, then $\mu'_W(m_{i-1}) >_{m_{i-1}} \mu_M(m_{i-1})$, for if not (m_{i-1}, w_i) would block μ'_W. On the other hand if $\mu'_W(m_i) >_{m_i} \mu_M(m_i)$ for some $i \in \{1, \ldots, k\}$, then $\mu_M(w_i) >_{w_i} \mu'_W(w_i)$, by stability of μ_M. The result follows by induction from the end of the chain, starting with the observation that w_{k+1} is single under μ'_W and matched under μ_M, so $\mu_M(w_{k+1}) >_{w_{k+1}} \mu'_W(w_{k+1})$.

Case 2: $\bar{M}$ has a man m_k who is single under μ_M. Let $(m_0, w_1, m_1, \ldots, w_k, m_k)$ be the path from m_0 to m_k. Hence $\mu'_W(m_i) = w_i$ for all $i = 1, \ldots, k$, and $\mu_M(m_i) = w_{i+1}$ for all $i = 0, \ldots, k-1$. We claim that the set $\{m_0, m_1, \ldots, m_k\}$ has the desired properties. Again all w_i, $i = 1, \ldots, k$ are in $\bar{W}$. Now if $\mu'_W(m_i) >_{m_i} \mu_M(m_i)$ for some $i \in \{1, \ldots, k\}$, then $\mu_M(w_i) >_{w_i} \mu'_W(w_i)$ by stability of μ_M, and $\mu'_W(m_{i-1}) >_{m_{i-1}} \mu_M(m_{i-1})$ by stability of μ'_W. The result follows by induction, since m_k is single under μ_M and matched under μ'_W to some woman in $\bar{W}$. Thus the proof is complete.

Remark: If more than one woman enters the market, let W' be the set of new women who are matched to men under μ'_W. For each woman $w_0 \in W'$ there exists a set $S(w_0)$ with the properties of Theorem 2.26. Thus we can obtain a set $S = \bigcup S(w_0)$, $w_0 \in W'$, such that S contains $\mu'_W(W')$ and all men in S and all women in $\mu_M(S)$ have the properties of Theorem 2.26. (Note that the only time there are no women in $\mu_M(S)$ is if $\mu_M(S) = S$, that is, if all men in S are single at μ_M.)

The next theorem takes another look at the sense in which the M-optimal stable matching, say, is optimal for the men. We have already studied the sense in which it is as good a *stable* matching as the men can achieve, but now we want to ask whether there might not be some other unstable matching that all the men would prefer. If so, then we might conclude that, even at the M-optimal stable matching, the men collectively "pay a price" for stability. However this turns out not to be the case – we will show there is no other matching, stable or not, that all men prefer to μ_M. (Such a matching is said to be "weakly Pareto optimal for the men," in honor of the Italian economist, Vilfredo Pareto, who first considered this kind of optimality.) By symmetry, of course, the parallel result holds for the women and μ_W.

Theorem 2.27: Weak Pareto optimality for the men.   *There is no individually rational matching μ (stable or not) such that $\mu >_m \mu_M$ for all m in M.*

We first give a short proof using the deferred acceptance algorithm with men proposing.

Proof: If μ were such a matching it would match every man m to some woman w who had rejected him in the algorithm in favor of some other man m' (i.e., even though m was acceptable to w). Hence all of these women, $\mu(M)$, would have been matched under μ_M. That is, $\mu_M(\mu(M)) = M$. Hence all of M would have been matched under μ_M and $\mu_M(M) = \mu(M)$. But since all of M are matched under μ_M, any woman who gets a proposal in the last step of the algorithm at which proposals were issued has not rejected any acceptable man, that is, the algorithm stops as soon as every woman in $\mu_M(M)$ has an acceptable proposal. So such a woman must be single at μ (since every man prefers μ to μ_M), which contradicts the fact that $\mu_M(M) = \mu(M)$.

Another proof, which follows, of Theorem 2.27 uses only the stability and optimality properties of μ_M. We will need first to introduce some terminology.

Definition 2.28.   *At a matching μ, m admires w if m and w are mutually acceptable and m prefers w to his mate $\mu(m)$.*

Note that m and w block μ if each admires the other.

Lemma 2.29 (Gale and Sotomayor).   *If no men are single at μ_M (i.e., if $\mu_M(M)$ is contained in W), then there is a woman w in $\mu_M(M)$ who has no admirers at μ_M.*

For the proof of Lemma 2.29 we need the following mathematical result about finite sets.

Lemma 2.30.   *Let f and g be functions from a finite set X into a set Y where f is a one-to-one correspondence. Then there is a nonempty subset A of X such that f and g map A onto $f(A)$.*

Proof: Since the function f is a one-to-one correspondence between X and Y we can define $h = f^{-1} \cdot g$. Then h maps X into X and since X is finite and $h^{n+1}(X)$ is contained in $h^n(X)$, we must have $h^k(X) = h^{k+1}(X)$ for some k. The set $A = h^k(X)$ has the desired property.

Proof of Lemma 2.29:   Suppose every w in $\mu_M(M)$ has an admirer and let $\alpha(w)$ be her favorite admirer. Applying Lemma 2.30 to the functions μ_M and α from $\mu_M(M)$ into M gives a set W' contained in $\mu_M(M)$, $W' \neq \phi$, such that $\alpha(W') = \mu_M(W')$, and $\alpha|_{W'}$ is one-to-one. Now define μ to agree with α on W', with α^{-1} on $\mu_M(W')$ and with μ_M otherwise. It is clear that μ is individually rational. Now μ is stable, for if m admires w, (m, w) does not block μ, since either $\mu(w) = \mu_M(w)$ and $\mu_M(w) >_w m$ by the stability of μ_M, or w is matched by μ to her favorite admirer whom she therefore prefers to m. But μ is preferred to μ_M by all m in $\mu(W')$, contradicting the fact that μ_M is M-optimal.

Alternate proof of Theorem 2.27:   If the conclusion were false, then every man would have to be matched under μ with some woman he admires under μ_M, so all of $\mu(M)$ would also have to be matched under μ_M (by strict preferences and the stability of μ_M), so $\mu_M(M) = \mu(M)$. From Lemma 2.29 there is at least one woman w in $\mu_M(M)$ who no one admires at μ_M. But this contradicts the conclusion that $w = \mu(m')$ for some m' who admires w, and so no such μ exists.

So the optimal stable matching for one side of the market is weakly Pareto optimal for the agents on that side of the market. This means there is no matching that all of those agents prefer. We might ask if we can strengthen the result to say that μ_M, for example, is *strongly* Pareto optimal for the men, which would mean there is no matching that all men like at least as well as μ_M and some men prefer. It turns out that this is not the case, and Example 2.31 illustrates a situation where μ_M is not strongly Pareto optimal.

Example 2.31 (Roth)
Let $M = \{m_1, m_2, m_3\}$ and $W = \{w_1, w_2, w_3\}$ with preferences over the acceptable people given by:

$$P(m_1) = w_2, w_1, w_3 \qquad P(w_1) = m_1, m_2, m_3$$
$$P(m_2) = w_1, w_2, w_3 \qquad P(w_2) = m_3, m_1, m_2$$
$$P(m_3) = w_1, w_2, w_3 \qquad P(w_3) = m_1, m_2, m_3.$$

Then,

$$\mu_M = \begin{array}{ccc} w_1 & w_2 & w_3 \\ m_1 & m_3 & m_2 \end{array}.$$

Nevertheless

$$\mu = \begin{array}{ccc} w_1 & w_2 & w_3 \\ m_3 & m_1 & m_2 \end{array}$$

leaves m_2 no worse than under μ_M, but benefits m_1 and m_3. So in general there may be matchings that all men like at least as well as the M-optimal stable matching, and that some men prefer. We shall return to this fact in our discussion of the strategic options available to coalitions of men.

2.5.1   *When preferences need not be strict*

Theorems 2.22 and 2.24–2.27 were all stated and proved under the assumption that preferences are strict. The conclusions of Theorem 2.22 no longer hold when agents may be indifferent between alternative mates: An agent may be single at one stable matching and married at another. It is easy to see this – consider a marriage market with two men and only one woman. If the woman prefers either man to being single, but is indifferent between them, and if both men prefer the woman to being single, then there are exactly two stable matchings, and each of the men is single at one of them and married at the other.

We have to be a little careful in discussing the conclusions of Theorems 2.24–2.27, since the statement of each of these theorems makes use of the fact that when preferences are strict, there exist optimal stable matchings, one for each side of the market. As we saw in Example 2.15, when preferences are not strict, optimal stable matchings may not exist.

Recall from our proof of Theorem 2.8, however, that the deferred acceptance algorithm, which produces an optimal stable matching when preferences are strict, continues to produce a stable matching when preferences are not strict, so long as the algorithm includes a way to break ties. The tie-breaking procedure can be thought of as any way of converting the true preferences P into a set of (complete and transitive) strict preferences P' that differ from P only in those comparisons in which indifference was expressed, so that the outcome of the algorithm with men proposing is the M-optimal stable matching with respect to P'. (Of course, two algorithms with different tie-breaking procedures may produce different matchings when preferences are not strict.) For any fixed tie-breaking procedure, we can now consider whether the conclusions of Theorems 2.24–2.27 hold with respect to the matching produced by the deferred acceptance algorithm with a tie-breaking procedure.

It is straightforward to see that the conclusions of Theorems 2.25 and 2.27 will continue to hold in this case, where μ_M and μ_W are now interpreted to mean the M- and W-optimal stable matchings under P'. (If the conclusions did not hold for some preferences P, then they would also fail to hold for the corresponding strict preferences P', and we have already proved that this cannot occur.)

The conclusions of Theorem 2.26 do not carry over unchanged to the case when preferences may not be strict. However the reason for this is obvious, since the theorem says that after a new woman is added to the market, some man will be better off. If the man in question is indifferent between his new spouse and his old one (which cannot occur when preferences are strict), then this will not be the case.

The conclusions of Theorem 2.24 fail to hold for a more interesting reason. If a man is indifferent between marrying some woman or remaining single, then when he extends his preferences to make another woman acceptable, she may appear higher in his preferences than a woman who was previously (barely) acceptable. The following example makes this clear.

Example 2.32: An example in which the conclusions of Theorem 2.24 fail to hold when preferences are not strict
Let the preferences of men $M = \{m_1, m_2\}$ and women $W = \{w_1, w_2, w_3\}$ be given by

$$P(m_1) = w_1, w_2 \qquad\qquad P(w_1) = m_1, m_2$$
$$P(m_2) = w_1, [w_2, m_2], w_3 \qquad P(w_2) = m_2$$
$$P(w_3) = m_2.$$

Then

$$\mu_W = \begin{matrix} m_1 & m_2 & (w_3) \\ w_1 & w_2 & w_3 \end{matrix}.$$

Now suppose that man m_2 extends his preferences so that his new preferences are given by $P(m_2) = w_1, w_3, [w_2, m_2]$. Then

$$\mu'_W = \begin{matrix} m_1 & m_2 & (w_2) \\ w_1 & w_3 & w_2 \end{matrix}.$$

Note that $\mu_W >_{w_2} \mu'_W$ but $\mu'_W >_{w_3} \mu_W$.

2.6   Guide to the literature

The game-theoretic analysis of marriage markets was begun by David Gale and Lloyd Shapley in their 1962 paper, in which Theorems 2.8 and 2.12 were proved. Little closely related work appeared until the 1972 paper of Shapley and Martin Shubik, which studied the class of related games called assignment games. Shapley and Shubik demonstrated results for that model parallel to Theorems 2.8, 2.12, and 2.16. (Although marriage markets and assignment games have some important differences, we will see in Chapter 8 that many of the conclusions that hold for one class of problems can also be shown to hold for the other.) For marriage markets, these latter two results were presented in the 1976 book of Donald Knuth, who attributes the lattice results to John Conway. Knuth's book, which is in French, also contains the decomposition lemma when $P = P'$ (Corollary 2.21). Some results concerning nonemptiness of the core in a family of generalizations of the assignment market that include the marriage market are found in Kaneko (1976, 1982, 1983).

Aside from Theorem 2.26, which is an adaptation and refinement made by the authors of a result presented for the assignment problem in Mo (1988a) (see Theorem 8.18), Section 2.5 largely follows the statement and proofs of Gale and Sotomayor (1985a, b). The first statement of Theorem 2.22 of which we are now aware appears in McVitie and Wilson (1970a) for the case when all men and women are mutually acceptable. It was also proved in Roth (1984a) in the context of the more general hospital-intern market that we will study in Chapter 5. Theorem 2.27 was proved, using the deferred acceptance procedure, in Roth (1982a), which also contains the related Example 2.31. (The statement of Lemma 2.29 used in our second proof of Theorem 2.27 corrects the original statement of that lemma in Gale and Sotomayor, 1985a). The conclusions of Theorem 2.25 were earlier reached, for a generalization of the assignment market, by Alexander Kelso and Vincent Crawford (1982). (We will have more to say about their work in Chapter 6.) A number of these conclusions were also reached by Demange and Gale (1985) for another generalization of the assignment market that we consider in Chapter 9. Results related to Theorem 2.25, presented in a non-game-theoretic context, can be found in earlier work on the assignment problem as a linear programming problem – see, for example, Shapley's 1962 paper. A result concerning the substitutability of players of the same type in a more general game theoretic context is contained in Scotchmer and Wooders (1989).

Gardenfors (1975) approaches the marriage problem from the perspective of voting theory. He defines a matching μ' to be "preferred by a majority" to another matching μ if the number of men and women who prefer μ' to μ is greater than the number who prefer μ to μ'. He then defines a matching μ to be a *majority assignment* if there exists no matching μ' that is preferred by a majority to μ. For the case that all men and women are acceptable to one another he observes that, when preferences are strict, every stable matching is a majority assignment. It is easy to see that this result also holds when agents are allowed to express a preference for remaining single.

Example 2.6 of a three-sided matching problem with no stable matchings comes from Alkan (1986). Some further exploration of one-sided matching, such as the roommate example considered in Example 2.5, are considered by Granot (1984), Gusfield (1988), and Irving (1986). Irving observes, among other things, that the task of finding stable matchings in the roommate problem is a generalization of the same task in the marriage problem, since the set of stable matchings is unchanged if a marriage problem is transformed into a roommate problem by appending to the very end of each agent's preference list all the other agents on the same side of the market. Bartholdi and Trick (1986) show that if agents in the roommate problem can be placed in a sequence with the property that each agent prefers a roommate who is closer to him or her in the sequence to one who is further away, then there is a unique stable matching. Hwang (1978) considers some questions concerning bounds on the number of preference profiles that produce a unique stable matching in the roommate problem (Hwang and Shyr (1977) consider similar questions for the marriage problem, and Hwang (1986) discusses some algebraic considerations associated with the number of stable matchings). Masarani and Gokturk (1988) consider a related question for the marriage problem, namely, the frequency of preferences such that there is a (unique) stable matching that can be reached by successively matching pairs of agents to their mutual first choices from those that remain on the market.

Although the primary motivation for our own consideration of the marriage model comes from problems that resemble labor markets, the possibility of using models of this sort to model marriage is explored at length by Becker (1981). He considers a much more highly structured model of the kind to be considered in Chapter 8, but in which preferences are not taken as primitives, but rather correspond in a given way to attributes of the agents (such as wage earning or homemaking ability). Thus although our results focus on things like the existence of stable outcomes for any preferences, his concern is with how the agents sort themselves out (e.g., high wage earners marry good cooks) for particular preferences.

As discussed earlier, an open question raised by Knuth (1976, problem 8) was whether, starting from an arbitrary matching, there always exists a path of blocking pairs leading to a stable matching. (Knuth's question referred to the problem in which there were equal numbers of men and women, all mutually acceptable. So in his problem all men and women were always matched. Here we speak of the related question in the more general model we consider.) As this book goes to press, this question has just been resolved in the affirmative by Roth and Vande Vate (1990). Formally, if (m', w') is a blocking pair for a matching μ, we say that a new matching ν is obtained from μ by *satisfying* the blocking pair if m' and w' are matched to one another at ν, their mates at μ (if any) are unmatched at ν (this differs from Knuth's formulation, in which any such "divorced" mates must marry one another), and all other agents are matched to the same mates at ν as they were at μ. We can state the following result.

Theorem 2.33 (Roth and Vande Vate).   *Let μ be an arbitrary matching for (M, W, P). Then there exists a finite sequence of matchings $\mu_1, \dots, \mu_k$, such that $\mu = \mu_1$, μ_k is stable, and for each $i = 1, \dots, k-1$, there is a blocking pair (m_i, w_i) for μ_i such that μ_{i+1} is obtained from μ_i by satisfying the blocking pair (m_i, w_i).*

The proof is not difficult, and can be sketched as follows. Start with an arbitrary matching μ, and select a subset S of agents such that there are no blocking pairs for μ contained in S, and μ does not match any agent in S to any agent not in S. (For example, S could be a pair of agents matched under μ, or a single agent.) A new player, say woman w, is selected to join S. If no man in S is part of a blocking pair with woman w, we may simply add her to S without changing the matching. Otherwise, select the man m whom woman w most prefers among those in S with whom she forms a blocking pair, and form a new matching by satisfying this blocking pair. If there is a woman $w' = \mu(m)$, then she is left unmatched at this new matching, and so there may now be a blocking pair (w', m') contained in S. If so, choose the blocking pair most preferred by w' to form the next new matching. The process continues in this way within the set $S \cup \{w\}$, much like the deferred acceptance algorithm with women proposing, satisfying the blocking pairs that arise at each step until the process terminates with a matching μ_i having no blocking pairs within $S_i = S \cup \{w\}$.

The process can now be continued, with the selected set S_i growing at each stage. At each stage, the selected set has no blocking pairs in it for the associated matching μ_i, and so the process converges to a stable matching when $S_k = M \cup W$.

Roth and Vande Vate go on to observe that an immediate corollary of the theorem is that a random process that begins from an arbitrary matching and continues by satisfying a randomly selected blocking pair must eventually converge with probability one to a stable matching, provided each blocking pair has a probability of being selected that is bounded away from zero. So the theorem gives an alternative proof of the nonemptiness of the set of stable matchings, and a family of alternative algorithms for reaching them.

CHAPTER 3

The structure of the set of stable matchings

This chapter discusses the mathematical structure of the set of stable matchings, including some computational algorithms. Some readers may prefer to simply skim this chapter, at least on first reading.

3.1   The core of a game

One of the most important "solution concepts" in cooperative game theory is the *core* of a cooperative game. In this section we will see that the set of stable matchings in a marriage problem is equal to the core of the game. (In subsequent chapters we will see that in some of the more complex two-sided matching markets we will consider, the set of stable matchings may be only a subset of the core, and in the case of many-to-many matching, [pairwise] stable matchings need not even be in the core at all.)

As discussed earlier, when we formally model various kinds of games, we will generally want to specify a set of *players,* a set of *feasible outcomes, preferences* of players over outcomes, and *rules,* which determine how the game is played. In how much detail we will want and need to specify the rules will depend on what kind of phenomena we are describing, and what kind of theory we are trying to construct. It is often sufficient to summarize the rules of the game by specifying which coalitions (i.e., subsets) of players are empowered by the rules of the game to enforce which outcomes. (Thus in our analysis of the marriage market in Chapter 2, we concentrated on the fact that in order for a marriage to take place, it is necessary and sufficient for the man and woman involved to agree.)

The rules of the game together with the specific preferences of the players induce a relation on the outcomes, called the *domination relation.*

Definition 3.1.   *For any two feasible outcomes x and y, x dominates y if and only if there exists a coalition of players S such that*

(a)   *every member of the coalition S prefers x to y; and*
(b)   *the rules of the game give the coalition S the power to enforce x (over y).*

Thus x dominates y if there is some coalition S whose members have both the incentive and the means to replace y with x. For this reason, if x dominates y, we might expect that y will not be the outcome of the game. This leads us to consider the set of undominated outcomes.

Definition 3.2.   *The core of a game is the set of undominated outcomes.*

Note that the difference between the definition of the core and the definition of the set of stable matchings for a marriage game lies in the fact that the core is defined via a domination relation in which *all* coalitions play a potential role, whereas the set of stable matchings is defined with respect to certain kinds of coalitions only. That is, an outcome fails to be in the core if it is "blocked" by any coalition of agents, whereas it fails to be a stable matching only if it is blocked by some individual agent or by some pair of agents consisting of a man and a woman.

   Formally, for a marriage market, a matching μ' dominates another matching μ if and only if there exists a coalition A contained in $M \cup W$, such that, for all men m and women w in A,

$$\mu'(m) \in A$$
$$\mu'(w) \in A$$
$$\mu'(m) \underset{m}{>} \mu(m)$$
$$\mu'(w) \underset{w}{>} \mu(w).$$

That is, the rules of the game allow the coalition A to enforce the set of marriages in the matching μ' that concerns its members if and only if every man in A is married to a woman in A and vice versa, and μ' dominates μ via such a coalition A if every member of A prefers μ' to μ. The following theorem shows that for the marriage market, nothing was lost by ignoring coalitions other than singletons and pairs.

Theorem 3.3.   *The core of the marriage market equals the set of stable matchings.*

Proof: If μ is individually irrational, then it is dominated via a singleton coalition, and if it is unstable via some man m and woman w, with $m >_w \mu(w)$ and $w >_m \mu(m)$, then it is dominated via the coalition $\{m, w\}$ by any matching μ' with $\mu'(m) = w$.

In the other direction, if μ is not in the core, then μ is dominated by some matching μ' via a coalition A. If μ is not individually irrational, this implies $\mu'(w) \in M$ for all w in A, since every woman w in A prefers $\mu'(w)$ to $\mu(w)$.

Let w be in A and $m = \mu'(w)$. Then m prefers w to $\mu(m)$ and μ is blocked by (m, w).

If a matching μ is not stable, then it is not in the core of the market; that is, there is some other matching μ' that dominates μ. If, say, μ is blocked by the pair (m, w), then μ is dominated by μ' with $\mu'(m) = w$. This does not imply that μ is dominated by an outcome that is itself in the core. The following theorem shows that in a marriage market, there is a sense in which this is *almost* the case. The proof is not so easy. (We will initially prove the theorem under the assumption that preferences are strict, and then show why this assumption is not needed.)

Theorem 3.4: Strong stability property (Demange, Gale, and Sotomayor).
If μ is an unstable matching, then either (a) there exist a blocking pair $\{m, w\}$ and a stable matching $\bar{\mu}$ such that $\bar{\mu}(m) \geq_m \mu(m)$ and $\bar{\mu}(w) \geq_w \mu(w)$; or (b) μ is not individually rational.

Note that the theorem does not say that the unstable matching μ is dominated by the stable matching $\bar{\mu}$. The result is not quite that strong, since the preferences in part (a) of the theorem need not be strict. But the theorem does say that there is a pair (m, w) such that $\bar{\mu}$ almost dominates μ via $\{m, w\}$, and that μ is in fact dominated by some μ' via the same coalition $\{m, w\}$ (since they are a blocking pair).

We will need the following lemma.

Lemma 3.5: Blocking lemma (Hwang; Gale and Sotomayor).   *Let μ be any individually rational matching with respect to strict preferences P and let M' be all men who prefer μ to μ_M. If M' is nonempty, there is a pair (m, w) that blocks μ such that m is in $M - M'$ and w is in $\mu(M')$.*

Proof of the blocking lemma

Case 1: $\mu(M') \neq \mu_M(M')$. Choose w in $\mu(M') - \mu_M(M')$, say, $w = \mu(m')$. Then m' prefers w to $\mu_M(m')$ so w prefers $\mu_M(w) = m$ to m'. But m is not in M' since w is not in $\mu_M(M')$, hence m prefers w to $\mu(m)$ (since preferences are assumed to be strict), so (m, w) blocks μ.

Case 2: $\mu_M(M') = \mu(M') = W'$. Let w be the last woman in W' to receive a proposal from an acceptable member of M' in the deferred acceptance algorithm. Since all w in W' have rejected acceptable men from M', w had some man m engaged when she received this last proposal. We claim (m, w) is the desired blocking pair. First, m is not in M' for if so, after having been rejected by w, he would have proposed again to a member of W', contradicting the fact that w received the last such proposal. But m prefers w to his mate under μ_M and since he is no better off under μ, he prefers w to $\mu(m)$. On the other hand, m was the last man to be rejected by w so she must have rejected her mate under μ before she rejected m and hence she prefers m to $\mu(w)$, so (m, w) blocks μ as claimed.

The following alternate proof of the blocking lemma shows how the result can be derived directly from considerations of stability and optimality, without using the deferred acceptance algorithm in the argument.

An alternate proof of the blocking lemma

Case 1:   Same proof as given.

Case 2: $\mu(M') = \mu_M(M') = W'$. Since $\mu >_{M'} \mu_M$, the stability of μ_M implies

$$\mu_M >_{W'} \mu. \tag{1}$$

Define a new market (M', W', P'). $P'(m)$ is the same as $P(m)$ restricted to $W' \cup \{m\}$ for all m in M'. For w in W', $P'(w)$ agrees with $P(w)$ restricted to $M' \cup \{w\}$ except that w is now ranked just below $\mu(w)$. In other words, the only men in M' who are acceptable to w are those m such that $m \geq_w \mu(w)$; so $\mu_M(w)$ is acceptable to w under P' for all w in W', by (1). Note that μ_M restricted to $M' \cup W'$ is still stable for (M', W', P'), because any pair that blocks μ_M under P' would also block it under P. Letting μ'_M be the M'-optimal matching for (M', W', P'), we see that

$$\mu'_M >_{M'} \mu_M. \tag{2}$$

That is, there is at least one m in M' who prefers μ'_M to μ_M because by hypothesis, $\mu >_m \mu_M$ for all m in M' and if $\mu_M = \mu'_M$ this would contradict the Pareto optimality theorem (Theorem 2.27). Furthermore,

$$\mu'_M \geq_{W'} \mu \tag{3}$$

by construction of P'. We now define μ' on $M \cup W$ by

$$\begin{aligned} \mu' &= \mu'_M \text{ on } M' \cup W' \\ &= \mu_M \text{ on } (M - M') \cup (W - W'). \end{aligned} \tag{4}$$

Since $\mu' >_M \mu_M$ by (2) and (4), we know that μ' is not stable for (M, W, P), so let $\{m, w\}$ be a blocking pair. We cannot have $\{m, w\}$ in $M' \cup W'$ because if so, m and w would be mutually acceptable under P', by construction of P', and so $\{m, w\}$ would block μ'_M. Further, if $m \in M'$ and $w \in W - W'$, then $\{m, w\}$ does not block μ' because then $\{m, w\}$ would block μ_M since m is no better off under μ_M than under μ'_M, by (2). If $m \in M - M'$ and $w \in W - W'$, $\{m, w\}$ does not block μ' because then it would block μ_M. Therefore we must have $m \in M - M'$ and $w \in W'$, but then $\{m, w\}$ also blocks μ, because w is at least as well off under μ'_M as under μ by (3). Hence $\{m, w\}$ is the desired blocking pair.

Note that the pair $\{m, w\}$ identified by the blocking lemma satisfies

$$\mu_M(m) \geq_m \mu(m) \quad \text{and} \quad \mu_M(w) \geq_w \mu(w).$$

The first statement just says m is in $M - M'$; as for the second, since w is in $\mu(M')$ we have $w = \mu(m') >_{m'} \mu_M(m')$ so $\mu_M(w) \geq_w \mu(w)$ by the stability of μ_M.

Proof of Theorem 3.4:   Assume μ is individually rational for the market (M, W, P). We need only consider the case where $\mu_M \geq_M \mu$, and symmetrically, $\mu_W \geq_W \mu$, since the blocking lemma gives the desired result otherwise with $\bar{\mu} = \mu_M$ or μ_W.

Now the set of stable matchings μ' such that $\mu' \geq_M \mu$ is nonempty since it contains μ_M, and it has a smallest element μ^* (since it is a lattice under the partial order $\geq_M$; see section 3.1.1). Since $\mu^* \geq_M \mu$, we can restrict our consideration to the case where

$$\mu^* \leq_W \mu \tag{1}$$

for if $\mu^*(w) >_w \mu(w)$ for some w, then the pair $(\mu^*(w), w)$ and the matching μ^* satisfy part (a) of the theorem.

We now define a new profile of preferences P' by modifying P as follows:

> Each w who is matched under the stable matchings deletes from her preference list of acceptable men all m such that $m <_w \mu^*(w)$. $\tag{2}$

> If $\mu^*(w) <_w \mu(w)$, then $\mu^*(w)$ is also deleted. $\tag{3}$

Clearly (3) must hold for some w for if not we would have $\mu = \mu^*$.

Now let μ'_M be the man-optimal stable matching for P'. We will show that μ'_M is the matching $\bar{\mu}$ of the theorem. First we claim μ'_M is stable under P. To see this, note that the woman-optimal stable matching μ_W is still individually rational for P' and hence also stable for P', since $\mu_W(w)$ is in $P'(w)$ for all w. Hence

$$\mu'_M \geq_M \mu_W. \tag{4}$$

Now if w is single under μ_M' she is single under μ_W and μ^* (Theorem 2.22). Hence from (2) and (3) w would have deleted no man from her preference list of acceptable men and hence, from (4) and the stability of μ_W, w could not be part of a blocking pair for μ_M' under P. On the other hand, if w is matched by μ_M', from (2) and (3) she prefers her mate to the men she has deleted. Hence she cannot block with any deleted man and hence she belongs to no blocking pair.

Next note that $\mu_M' \leq_M \mu^*$, for if $\mu_M'(m) >_m \mu^*(m)$ and $w = \mu_M'(m)$, then $\mu^*(w) >_w m$ by the stability of μ^*, but this means from (2) that m was deleted by w so $w = \mu_M'(m)$ is impossible.

It follows that $\mu(m) >_m \mu_M'(m)$ for at least one m, for if not we would have $\mu \leq_M \mu_M' \leq_M \mu^*$, but since from (3) $\mu_M' \neq \mu^*$ (recall that $\mu \neq \mu^*$), this would imply $\mu(m) \leq_m \mu_M'(m) <_m \mu^*(m)$ for at least one m. But μ^* is the smallest stable matching preferred by M to μ, which is a contradiction.

Finally, we apply the blocking lemma to the preferences profile P' for which μ_M' is man-optimal. Then there is a blocking pair for μ under P' and hence under P and the proof is complete with $\bar{\mu} = \mu_M'$ as claimed, under the assumption that preferences are strict.

To prove the theorem without the assumption that preferences are strict, we need the following additional observation. Let μ be an unstable matching under nonstrict preferences P. Then there exists a way to break ties so that the strict preferences P' correspond to P, and every pair (m, w) that blocks μ under P' also blocks μ under P. (If any agent x is indifferent under P between $\mu(x)$ and some other alternative, then under P', x prefers $\mu(x)$.) Then the theorem applied to the case of the strict preferences P' gives the desired result.

Note that if μ is not individually rational, someone, say a man m, is matched under μ with a woman not on his list of acceptable women, so $\bar{\mu}(m) >_m \mu(m)$ is true for any stable $\bar{\mu}$.

3.1.1 *Lattices and the algebraic structure of the set of stable matchings*

Consider a set L endowed with some partial order $\geq$. If a and b are in L and $a \geq b$ we will say that a is "greater than or equal to" b. Analogously, if $a \leq b$ we will say that a is "less than or equal to" b. An *upper bound* of a subset X of L is an element a in L such that $a \geq x$ for all x in X. Denote by $\sup X$ the least upper bound of X if this upper bound exists. By the antisymmetry property of "$\geq$" ($a \geq b$ and $b \geq a$ implies $a = b$), it follows that $\sup X$ is unique if it exists. A *lower bound* of X is defined dually. Denote by $\inf X$ the greatest lower bound of X if this lower bound exists. Again, by the antisymmetry property of "≥", inf X is unique if it exists.

Definition 3.6. *A **lattice** is a partially ordered set L any two of whose elements x and y have a "sup", denoted by $x \vee y$ and an "inf", denoted by $x \wedge y$. A lattice L is **complete** when each of its subsets X has a "sup" and an "inf" in L.*

Hence, any nonempty complete lattice L has a least element $\underline{x}$ and a greatest element $\bar{x}$.

The binary operations $\wedge$ and $\vee$ in lattices have some analogous properties to those of ordinary multiplication and addition. In many lattices, this analogy includes the distributive law $x(y+z) = xy + xz$.

Definition 3.7. *A lattice L is **distributive** if and only if*

$$x \wedge (y \vee z) = (x \wedge y) \vee (x \wedge z) \quad and$$

$$x \vee (y \wedge z) = (x \vee y) \wedge (x \vee z), \text{ for all } x, y \text{ and } z \text{ in } L.$$

We showed in Theorem 2.16 that the set of stable matchings is a lattice. It is left to the reader to verify Theorem 3.8.

Theorem 3.8 (Conway). *When preferences are strict, the set of stable matchings is a distributive lattice under the common order of the men, dual to the common order of the women.*

As we saw when we initially discussed the lattice theorem, the algebraic structure of the set of stable matchings gives us some insight into the conflict and coincidence among agents in these markets. We might therefore hope to say something more about what kind of lattices arise as sets of stable matchings, in order to use any additional properties thus specified to learn more about the market. The following theorem shows that this line of investigation will not bear any further fruit. We state it here without proof.

Theorem 3.9 (Blair). *Every finite distributive lattice equals the set of stable matchings of some marriage market.*

3.2   Computational questions

In this section we turn to some computational questions about the marriage model. To keep things relatively simple we confine our attention to the case when all preferences are strict.

In Section 3.2.1 we describe an algorithm, due to Irving and Leather, to compute every stable matching in the marriage problem. To describe the algorithm it will be necessary to explore some properties of the set of stable matchings that are interesting in their own right. In order to enumerate the stable matchings, we will want to consider how "consecutive" stable matchings in the lattice of stable matchings are related, where two stable matchings μ and μ' are called consecutive if $\mu >_M \mu'$ and if there is no stable matching μ'' such that $\mu >_M \mu'' >_M \mu'$. It turns out that the men who like μ' less than μ form a cycle $a_1, \ldots, a_r$ such that $\mu'(a_i) = \mu(a_{i+1})$ for $i = 1, \ldots, r$ (and $a_{r+1} = a_1$).

It will be easiest to describe such cycles in terms of certain "reduced" preference lists obtained from the original preference lists by eliminating some unachievable mates. The first part of Section 3.2.1 is devoted to describing these lists and their properties. The algorithm for finding all stable matchings then proceeds by starting at μ_M and generating further stable matchings by branching out along all branches of the lattice (by identifying the cycle needed to reach each consecutive matching) until μ_W has been reached.

In Section 3.2.2 we turn to a related question, How many stable matchings may there be in a marriage market with n men and n women? We will see that the number of stable matchings can grow exponentially with the number of participants in the market. This means that, for markets with large numbers of participants, it may not be feasible to actually compute each stable matching.

However in Section 3.2.3 we will see that the problem of determining all pairs (m, w) of achievable mates is simpler, since any path through the lattice connecting the M- and W-optimal stable matchings will encounter all such pairs.

We also discuss some very recent results concerning the linear structure of the set of stable matchings in Section 3.2.4.

3.2.1 *An algorithm to compute every stable matching*

In this section if P is a preference profile then $P(x)$ will mean agent x's list of acceptable people with the inclusion of x as the last entry.

It follows from the optimality of μ_M and μ_W that if (m, w) is a mutually acceptable pair and $w >_m \mu_M(m)$, or $m >_w \mu_W(w)$, then m and w cannot be matched to each other at any stable matching. This suggests that the preference lists (of acceptable spouses) can be shortened by the following "reduction" procedure, without changing the set of stable matchings. For all m in M and all w in W:

Step 1: Remove from m's list of acceptable women all w who are more preferred than $\mu_M(m)$. Remove from w's list of acceptable men all m who are more preferred than $\mu_W(w)$.

Therefore, $\mu_M(m)$ will be the first entry in m's reduced list and $\mu_W(w)$ will be the first entry in w's reduced list.

Since μ_M gives to any woman her worst achievable mate (and μ_W gives to any man his worst achievable mate), and if m and w are matched to each other under some stable matching, then (m, w) is a mutually acceptable pair, we can proceed as follows:

Step 2:   Remove from w's list of acceptable men all m who are less preferred than $\mu_M(w)$. Remove from m's list of acceptable women all w who are less preferred than $\mu_W(m)$.

Thus, $\mu_M(w)$ will be the last entry in w's reduced list and $\mu_W(m)$ will be the last entry in m's reduced list.

Step 3:   After steps 1 and 2, if m is not acceptable to w (i.e., if m is not on w's preference list as now modified), then remove w from m's list of acceptable women, and similarly remove from w's list of acceptable men any man m to whom w is no longer acceptable.

Hence, m will be acceptable to w if and only if w is acceptable to m after step 3.

In general, if μ is any stable matching and we replace μ_M by μ in the reduction process described by steps 1–3, the resulting profile will be called a *profile of reduced lists* for the original market and will be denoted by $P(\mu)$.

It is clear from the construction of $P(\mu)$ that

(1) $\mu(m)$ is the first entry of $P(\mu)(m)$ and $\mu(w)$ is the last entry of $P(\mu)(w)$. $\mu_W(m)$ is the last entry of $P(\mu)(m)$ and $\mu_W(w)$ is the first entry of $P(\mu)(w)$.

(2) μ is the M-optimal stable matching under $P(\mu)$ and μ_W is the W-optimal stable matching under $P(\mu)$.

(3) m is acceptable to w if and only if w is acceptable to m under $P(\mu)$.

(4)   If a man or woman is the only acceptable man or woman for someone on the other side of the market, then he or she is not acceptable to anyone else.

Furthermore, if m and w are mutually acceptable under the original preferences but they are not in each other's lists under $P(\mu)$, then either

(5) $w >_m \mu(m)$ or $\mu(w) >_w m$ or $\mu_W(m) >_m w$ or $m >_w \mu_W(w)$. So, (recall Theorem 2.13)

(6)   If μ' is stable under the original preferences and $\mu \geq_M \mu'$, then $\mu'(m) \in P(\mu)(m)$ and $\mu'(w) \in P(\mu)(w)$.

Proposition 3.10. *If μ' is a matching then μ' is stable under $P(\mu)$ if and only if μ' is stable under the original preferences and $\mu \geq_M \mu'$.*

Proof: If μ' is stable under $P(\mu)$ then $\mu \geq_M \mu'$ by (1). Now if (m, w) blocks μ' under the original preferences, then $m >_w \mu'(w) \geq_w \mu(w)$, since $\mu'(w) \in P(\mu)(w)$ and $\mu(w)$ is the least preferred under $P(\mu)(w)$. So $m >_w \mu(w)$, from which it follows that $\mu(m) >_m w$, for if not (m, w) would block μ. But then using (5), w is acceptable to m under $P(\mu)(m)$ and m is acceptable to w under $P(\mu)(w)$, which means that (m, w) blocks μ' under $P(\mu)$, which is a contradiction. The other direction follows immediately from (6).

Proposition 3.11. *Let $P(\mu)$ be a profile of reduced lists for $(M, W; P)$. If $P(\mu)(m)$ contains at most one acceptable woman for all m in M, then μ is the W-optimal stable matching for $(M, W; P)$.*

Proof: Immediate from the fact that $\mu_W(m)$ is the last entry in $P(\mu)(m)$ for all m in M.

Definition 3.12. *A set of men $\{a_1, \ldots, a_r\}$ defines a cycle for some profile of reduced lists, $P(\mu)$, if*

(i) *for $i = 1, \ldots, r - 1$, the second woman in $P(\mu)(a_i)$ is $\mu(a_{i+1})$ (i.e., the first woman in $P(\mu)(a_{i+1})$).*

(ii) *the second woman in $P(\mu)(a_r)$ is $\mu(a_1)$, (i.e., the first woman in $P(\mu)(a_1)$).*

We denote such a cycle by $\sigma = (a_1, \ldots, a_r)$ and we say that a_i generates the cycle σ, for any $i = 1, \ldots, r$. (Think of each man a_i in the cycle as being asked to point to the man presently matched to man a_i's next choice, after $\mu(a_i)$.)

It is clear that we can only have a cycle if $P(\mu)(m)$ has more than one acceptable woman for some m. In this case it is very easy to find a cycle. Let p_1 be an arbitrary man such that $P(\mu)(p_1)$ contains more than one woman. Then construct an oriented graph, whose nodes are $M \cup W$, as follows.

There is an arc from p_i to q_{i+1} if q_{i+1} is the second woman in $P(\mu)(p_i)$.

There is an arc from q_i to p_i if p_i is the last man in $P(\mu)(q_i)$ [that is, $\mu(q_i) = p_i$].

This graph will close at some step (when some q_s or p_s is repeated) and then we will get a cycle. To see this, imagine instead that the graph stops at a not previously reached q_i. Then q_i is single under μ, and her list of acceptable men under $P(\mu)$ is empty. But q_i was reached from p_{i-1}, which from the construction of the reduced lists means that $p_{i-1} \in P(\mu)(q_i)$, which is the necesssary contradiction.

Similarly, if the graph were to stop at some not previously reached p_i, then there is no second woman in $P(\mu)(p_i)$. Then $\mu(p_i) = \mu_W(p_i) = q_i$ and the only acceptable man in $P(\mu)(q_i)$ is p_i. So $q_i \notin P(\mu)(p_{i-1})$ and q_i could not have been reached from p_{i-1}.

We have shown the following:

Proposition 3.13. *Let $P(\mu)$ be a profile of reduced lists for (M, W, P). There is a cycle for $P(\mu)$ if and only if $P(\mu)(m)$ has more than one acceptable woman for some m.*

If $P(\mu)(m)$ has more than one acceptable woman, the proposition asserts there is some man who generates a cycle. As the following example shows, this man is not necessarily m: $P(\mu)(m_3)$ contains two women but m_3 does not generate a cycle. The only cycle for $P(\mu)$ is $\sigma = (m_1, m_2)$.

$$P(m_1) = w_1, w_2, w_3 \qquad P(w_1) = m_2, m_1$$
$$P(m_2) = w_2, w_1 \qquad P(w_2) = m_3, m_1, m_2$$
$$P(m_3) = w_3, w_2 \qquad P(w_3) = m_1, m_3$$

Suppose now that μ is a stable matching and $\sigma = (a_1, \ldots, a_r)$ is a cycle for $P(\mu)$. Then we can define a matching μ' as follows:

$$\mu'(a_i) = \mu(a_{i+1}), \quad \text{for all } i = 1, \ldots, r-1$$
$$\mu'(a_r) = \mu(a_1)$$
$$\mu'(m) = \mu(m), \quad \text{for all } m \text{ not in } \sigma.$$

We will refer to μ' as a *cyclic matching under $P(\mu)$*.

Proposition 3.14. *Let $P(\mu)$ be a profile of reduced lists. If μ' is a cyclic matching under $P(\mu)$, then μ' is stable under the original preferences.*

Proof: By Proposition 3.10 it is enough to show that μ' is stable under $P(\mu)$. Suppose instead there is a blocking pair (m, w). Let σ be the cycle associated with μ'. Then $m \in \sigma$, otherwise m is matched to his first choice under μ' and (m, w) would not be a blocking pair. Then $w = \mu(m)$, since $w >_m \mu'(m)$ and $\mu'(m)$ is m's second choice under $P(\mu)(m)$. But w prefers $\mu'(w)$ to m, because m is her last choice, which is the necessary contradiction.

The algorithm can now be stated:

Step 1: Find μ_M and μ_W (by the deferred acceptance procedures), and $P(\mu_M)$.

Step k: For each profile P' of reduced lists obtained in step $(k-1)$, find all corresponding cycles and for each cycle obtain the corresponding cyclic matching for P'. Then for each cyclic matching μ, obtain the profile of reduced lists $P(\mu)$.

It is clear that this algorithm stops after a finite number of steps. We have:

Proposition 3.15. *The algorithm stops at step t if and only if we obtain only one profile of reduced lists in this step and the men's lists of acceptable women have at most one woman.*

Proof: If the algorithm stops at step t, this means that every profile of reduced lists obtained at this step has no cycles, for if not t would not be the last step. Then they have at most one acceptable woman in the men's lists, by Proposition 3.13. On the other hand, it follows from Proposition 3.11 that the M-optimal matching under each one of these profiles must be the W-optimal matching under the original preferences, which shows that all these profiles must be the same.

Lemma 3.16. *Let $P(\mu)$ be a profile of reduced lists for (M, W, P) and μ' be a stable matching for $P(\mu)$. If $\mu \neq \mu'$, then there exists a cyclic matching under $P(\mu)$, μ'', such that $\mu'' \geq_M \mu'$.*

Proof: If $\mu \neq \mu'$, then there exists some m' in M such that $\mu(m') >_{m'} \mu'(m') \geq_{m'} m'$, since μ is the M-optimal matching for $P(\mu)$. By Theorem 2.22 $\mu'(m')$ is in W and so $P(\mu)(m')$ has more than one acceptable woman. By Proposition 3.13 we can find a cycle σ for $P(\mu)$ by generating the sequence:

$$p_1 = m', \qquad q_1 = \mu(p_1).$$

For all $i > 1$,

$$q_i = 2\text{nd acceptable woman in } P(\mu)(p_{i-1})$$

$$p_i = \mu(q_i).$$

Since $\mu''(m) = \mu(m) \geq_m \mu'(m)$ for all $m \notin \sigma$, we only need to show that $\mu''(m) \geq_m \mu'(m)$ for all $m \in \sigma$, which is equivalent to showing that $\mu(m) >_m \mu'(m)$ for all m in σ. We are going to show this holds for all p_i in the sequence. Suppose not. Let p_i be the first man in the sequence such that $q_i = \mu(p_i) = \mu'(p_i)$. Then $p_i \neq p_1$ and q_i prefers p_{i-1} to p_i, by definition of p_i. From the stability of μ' it follows that p_{i-1} must prefer $\mu'(p_{i-1})$ to q_i. But q_i is the second entry of $P(\mu)(p_{i-1})$, so $\mu'(p_{i-1}) = \mu(p_{i-1})$, which contradicts the assumption that p_i is the first man in the sequence to be matched to his first choice under μ'. Then $\mu(p_i) >_{p_i} \mu'(p_i)$ for all p_i in the sequence and in particular for all m in σ.

Theorem 3.17. *If μ is a stable matching under the original preferences and $\mu \neq \mu_M$, then μ is some cyclic matching under some profile of reduced lists obtained in some step of the algorithm.*

Proof: Suppose μ is not a cyclic matching and $\mu_M \neq \mu$. So there is a cyclic matching μ_1, under $P(\mu_M)$, such that $\mu_M >_M \mu_1 >_M \mu$, by Lemma 3.16. Then μ is stable under $P(\mu_1)$ by Proposition 3.10, and we can apply Lemma 3.16 again to get a cyclic matching μ_2, under $P(\mu_1)$, such that $\mu_1 >_M \mu_2 >_M \mu$, and so on. Since we have a finite number of stable matchings, the sequence $\mu_M >_M \mu_1 >_M \mu_2 >_M \cdots$ has a smallest element, μ_k, and $\mu_k >_M \mu$. But then we can use the same reasoning as before and get a smaller cyclic matching μ_{k+1}, which is a contradiction.

Proposition 3.14, Lemma 3.16, and Theorem 3.17 imply that the matchings obtained at each step of the algorithm from any reduced lists $P(\mu)$ are the stable matchings consecutive to μ in the lattice of stable matchings. That is, if μ and μ' are consecutive stable matchings, then there is a cycle σ formed by the subset of men who prefer μ to μ', and every man is matched by μ' to his second choice under $P(\mu)$. (See also the proof of Theorem 3.20.) Note that a given stable matching may be computed more than once in the course of the algorithm. For example, in Example 2.17, μ_4 is computed as a cyclic matching from both $P(\mu_2)$ and $P(\mu_3)$ (see the figure accompanying that example). We have presented the algorithm in this way for clarity; obviously to optimize it for computational purposes much of this double counting could be eliminated.

3.2.2 *The number of stable matchings*

In this section we want to get some idea of how the maximum possible number of stable matchings may grow as a function of the number of men and women. For this purpose we define an *instance of size n* to be a marriage market (M, W, P_n) with n men and n women. As n increases, the maximum number of stable matchings grows exponentially. Theorem 3.19 shows this for when n is a power of two.

Lemma 3.18 (Irving and Leather). *Consider an instance (M, W, P_n) of size n of the stable marriage problem with $g(n)$ stable matchings, and such that every pair (m, w) is mutually acceptable. Then there exists an instance (M^*, W^*, P_{2n}^*) of size $2n$ with at least $2[g(n)]^2$ stable matchings.*

Proof: Let (M, W, P_n) have $M = \{m_1, \ldots, m_n\}$ and $W = \{w_1, \ldots, w_n\}$. Construct an *isomorphic* instance (M', W', P_n') with $M' = \{m_{n+1}, \ldots, m_{2n}\}$ and $W' = \{w_{n+1}, \ldots, w_{2n}\}$, such that if $w_j >_{m_i} w_k$ for some $i \le n$, then $w_{j+n} >_{m_{i+n}} w_{k+n}$, and if $m_j >_{w_i} m_k$ for some $i \le n$, then $m_{j+n} >_{w_{i+n}} m_{k+n}$. Then both markets have the same number of stable matchings. Now construct an instance (M^*, W^*, P_{2n}^*) of size $2n$ in the following way. $M^* = \{m_1, \ldots, m_{2n}\}$, $W^* = \{w_1, \ldots, w_{2n}\}$ and for $i \le n$, the m_i's (respectively m_{i+n}'s) preference list is *followed* by the m_{i+n}'s (respectively m_i's) preference list. For $i \le n$, the w_i's (respectively w_{i+n}'s) preference list is *preceded* by the w_{i+n}'s (respectively w_i's) preference list.

Now, let μ and μ' be stable matchings for (M, W, P_n) and (M', W', P_n'), respectively. Since every pair (m, w) is mutually acceptable, no man or woman is single at μ or μ'. Let μ_1 be a matching for (M^*, W^*, P_{2n}^*) defined as follows:

$$\mu_1(m) = \mu(m) \quad \text{if } m \in M$$

$$\mu_1(m) = \mu'(m) \quad \text{if } m \in M'.$$

Then μ_1 is stable since no (m, w) in $M \times W$ (respectively in $M' \times W'$) can block μ_1 because of the stability of μ (respectively μ'). Furthermore, no m in M and w in W', or m in M' and w in W can cause instability because such a man prefers his partner to any such woman.

By similar argument we can have another stable matching μ_2, for (M^*, W^*, P_{2n}^*).

$$\mu_2(w_i) = \mu'(w_{i+n}) \quad \text{if } i \le n$$

$$\mu_2(w_i) = \mu(w_{i-n}) \quad \text{if } i \ge n.$$

Hence, as μ and μ' run independently through all $g(n)$ stable matchings of (M, W, P_n) and (M, W, P_n'), we obtain $2\{g(n)\}^2$ stable matchings for (M^*, W^*, P_{2n}^*).

Theorem 3.19 (Irving and Leather). *For each $i \ge 0$ there exists an instance (M, W, P_n) of size $n = 2^i$ with at least 2^{n-1} stable matchings.*

Proof: Such a sequence of marriage markets is obtained by starting with the only marriage market of size 1, which has of course a single stable matching, and repeatedly applying the construction of Lemma 3.18. If $g(n)$ denotes the number of stable matchings of the marriage market of size n so generated, then we have

$$g(n) \geq 2\{g(n/2)\}^2, \quad g(1) = 1$$

from which the result of the Theorem follows.

Note that the example due to Knuth used to illustrate the lattice theorem (Example 2.17) was constructed along the lines indicated in Lemma 3.18.

3.2.3   Finding all achievable pairs

In this section we give a method for finding all achievable pairs, that is, all pairs (m, w) such that m is matched to w at some stable matching. The algorithm can be stated as follows:

Step 1: Find μ_M and μ_W, by the deferred acceptance procedure, and $P(\mu_M)$.

Step k: For each profile P' of reduced lists obtained in step $(k-1)$, find one corresponding cycle (if none exists, stop) and obtain the corresponding cyclic matching μ for P'. Then obtain the profile of reduced lists $P(\mu)$.

Observe that in each step (other than the last) of this algorithm we obtain exactly *one* cyclic matching, whereas in the previous algorithm we might obtain more than one.

It is clear that this algorithm stops after a finite number of steps and Proposition 3.15 applies. By Propositions 3.11 and 3.13 the W-optimal stable matching is obtained at the last step. Now we will prove that by this algorithm we obtain all achievable pairs.

Theorem 3.20 (Gusfield).   *Let* $\mu_M = \mu_0 >_M \mu_1 >_M \mu_2 >_M \cdots >_M \mu_t = \mu_W$ *be the sequence of stable matchings obtained by the algorithm just described. Then every achievable pair appears in at least one of the matchings in the sequence.*

Proof: There is no stable matching μ between μ_i and μ_{i+1} for all $i = 0, \ldots, t-1$. (Suppose this is false. Then there would exist some stable matching μ and some m such that $\mu_i(m) >_m \mu >_m \mu_{i+1}(m)$ and $\mu_i >_M \mu >_M \mu_{i+1}$, for some $i = 0, \ldots, t-1$. By Proposition 3.10 μ is stable under $P(\mu_i)$, so $\mu(m)$ is acceptable to m under $P(\mu_i)$. But then $\mu_{i+1}(m)$ is neither m's first choice nor m's second choice under $P(\mu_i)$, which contradicts the definition of μ_{i+1}.)

Now let μ_i and μ_{i+1} be two consecutive stable matchings in the sequence, and let m be such that $\mu_i(m) = w_i \neq w_{i+1} = \mu_{i+1}(m)$. Of course man m prefers w_i to w_{i+1}. Now let w be a woman such that $w_i >_m w >_m w_{i+1}$. If there exists a stable matching μ such that $\mu(m) = w$, then $\mu' = \min\{\mu_i, \max\{\mu, \mu_{i+1}\}\}$ with respect to $>_M$ (i.e., $\mu' = \mu_i \wedge \{\mu \vee \mu_{i+1}\}$) is also a stable matching in which $\mu'(m) = w$. Hence μ' is different from both μ_i and μ_{i+1} and then $\mu_i >_M \mu' >_M \mu_{i+1}$, which contradicts that there is no stable matching between μ_i and μ_{i+1}.

The theorem says the following. Consider the diagram of the lattice of all stable matchings under $>_M$. There is an arc from μ to μ' if $\mu >_M \mu'$ and there is no stable matching between μ and μ'. Then the matchings along any path between μ_M and μ_W contain all the achievable pairs.

Figure 3.1 shows again the diagram of the lattice of all stable matchings for the marriage problem given by Knuth in Example 2.17. The preceding remark is easy to verify. The number in position i indicates the woman matched to man i in that matching.

3.2.4 *The linear structure of the set of stable matchings*

This section demonstrates the surprising result that the set of stable matchings can be represented as the extreme points of the set of solutions of a simple system of *linear* constraints.

For simplicity we consider the special case in which $|M| = |W|$ and every pair (m, w) is mutually acceptable, and all preferences are strict. Thus, every man is matched to some woman and vice versa, under any stable matching. Let the *configuration* of a matching μ be a matrix x of zeros and ones such that $x_{mw} = 1$ if $\mu(m) = w$ and $x_{mw} = 0$ otherwise. We will sometimes simply call x itself a matching if it is the configuration of some matching μ.

In what follows we will also consider matrices x of dimension $|M| \times |W|$ whose elements may not be integers, that is, matrices that may not be the configuration of any matching. We will use the notation $\sum_i x_{iw}$ to denote the sum over all i in M, $\sum_j x_{mj}$ to denote the sum over all j in W, $\sum_{j >_m w} x_{mj}$ to denote the sum over all those j in W that man m prefers to woman w, and $\sum_{i >_w m} x_{iw}$ to denote the sum over all those i in M that woman w prefers to man m.

With this notation, we can characterize the stable matchings by their configurations as follows.

Theorem 3.21 (Vande Vate). *A matching is stable if and only if its configuration x is an **integer** matrix of dimension $|M| \times |W|$ satisfying the following set of constraints:*

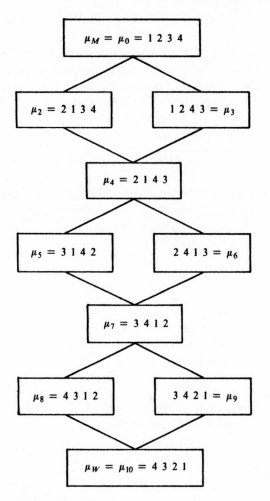

Figure 3.1.

(1) $\sum_j x_{mj} = 1$ *for all m in M,*

(2) $\sum_i x_{iw} = 1$ *for all w in W,*

(3) $\sum_{j >_m w} x_{mj} + \sum_{i >_w m} x_{iw} + x_{mw} \geq 1$ *for all m in M and w in W,*   *and*

(4) $x_{mw} \geq 0$ *for all m in M and w in W.*

Constraints (1), (2), and (4) require that if x is integer it is a matching, that is, its elements are 0's and 1's and every agent on one side is matched to some agent on the opposite side. It is easy to check that constraint (3) is equivalent to the nonexistence of blocking pairs. (To see this, note that if x is a matching, i.e., a matrix of 0's and 1's satisfying (1), (2), and (4), then (3) is not satisfied for some m and w only if $\sum_{j >_m w} x_{mj} = \sum_{i >_w m} x_{iw} = x_{mw} = 0$, in which case m and w form a blocking pair.)

Thus an integer $|M| \times |W|$ matrix x is the configuration of a stable matching if and only if x satisfies (1)–(4). Of course there will in general be an infinite set of noninteger solutions of (1)–(4) also, and these are not matchings. However we may think of them as "fractional matchings," in which x_{mw} denotes something like the fraction of the time man m and woman w are matched, or the probability that they will be matched.

An *extreme point* of a convex set C is a point x in C for which there exist no y and z in C, both distinct from x, such that $x = \alpha y + (1-\alpha)z$ for some $0 < \alpha < 1$.

The surprising result is that the integer solutions of (1)–(4), that is, the stable matchings, are precisely the extreme points of the convex polyhedron defined by the linear constraints (1)–(4). That is, we have the following result.

Theorem 3.22 (Vande Vate). *Let C be the convex polyhedron of solutions to the linear constraints (1)–(4). Then the integer points of C are precisely its extreme points. That is, the extreme points of the linear constraints (1)–(4) correspond precisely to the stable matchings.*

It is well known (as Birkhoff's theorem) that integer solutions of constraints (1), (2), and (4) are extreme points of the set defined by those three constraints. (Consider an integer solution x and suppose $x = \alpha y + (1-\alpha)z$ for some $0 < \alpha < 1$. Then each x_{ij} equals 0 or 1, so $y_{ij} = z_{ij} = x_{ij}$ for each i, j in $M \times W$, so $y = z = x$, and x is extreme.) The set C of solutions to (1)–(4) is a subset of the solutions to (1), (2), and (4), so its integer solutions are extreme points of C, since they are extreme points of the larger set. What remains to be shown is that every extreme point of C is integer, that is, is a matrix of 0's and 1's.

The proof that follows is due to Rothblum (the original proof of Vande Vate was more complex).

Let x be any matrix (not necessarily integer) that satisfies (1)–(4). For any m in M define $\bar{x}(m)$ and $\underline{x}(m)$ respectively as m's most preferred and least preferred elements of the set $\{w$ in W such that $x_{mw} > 0\}$. Similarly, for each w in W define $\bar{x}(w)$ and $\underline{x}(w)$ as the most and least preferred (with respect to w's preferences $>_w$) of the men with whom w is matched with positive weight by x.

Note that x is integer if and only if $\bar{x}(m) = \underline{x}(m)$ for all m in M, or equivalently, if and only if $\bar{x}(w) = \underline{x}(w)$ for all w in W.

The following two lemmas about fractional matchings are reminiscent of the decomposition lemma.

Lemma 3.23 (Rothblum).   *Let x be an $|M| \times |W|$ matrix satisfying (1)–(4). Then*

(5) $w \geq_m \bar{x}(m)$ *implies* $m \leq_w \underline{x}(w)$,   *and*

(6) $w = \bar{x}(m)$ *if and only if* $m = \underline{x}(w)$.

Furthermore, if

(7) $\sum\limits_{j >_m w} x_{mj} + \sum\limits_{i >_w m} x_{iw} + x_{mw} = 1,$

then the reverse implication of (5) holds also.

Since the marriage market we are considering is symmetric between men and women, once we have proved the preceding result we will also have the symmetric result, obtained by exchanging the roles of M and W, as follows.   .

Lemma 3.24 (Rothblum).   *Let x be an $|M| \times |W|$ matrix satisfying (1)–(4). Then*

(8) $m \geq_w \bar{x}(w)$ *implies* $w \leq_m \underline{x}(m)$,   *and*

(9) $m = \bar{x}(w)$ *if and only if* $w = \underline{x}(m)$.

Furthermore, if (7) holds then the reverse implication of (8) holds also.

Proof of Lemma 3.23:   If $w \geq_m \bar{x}(m)$ then $x_{mj} = 0$ for all $j >_m w$, and so $\sum_{j >_m w} x_{mj} = 0$. So (3) implies $\sum_{i >_w m} x_{iw} + x_{mw} \geq 1$, and (2) and (4) therefore imply $\sum_{i <_w m} x_{iw} = 0$, which proves (5).

To prove (6) in one direction, suppose $w = \bar{x}(m)$. Then $x_{mw} > 0$ and so $m \geq_w \underline{x}(w)$. But $m \leq_w \underline{x}(w)$ by (5), and since preferences are strict this implies $m = \underline{x}(w)$ as required.

Note that the map $\bar{x}: M \to W$ is one-to-one, since if $w = \bar{x}(m_1) = \bar{x}(m_2)$ then $m_1 = m_2 = \underline{x}(w)$. Since $|M| = |W|$, the map is also onto.

So to prove (6) in the other direction, suppose $m = \underline{x}(w)$. Then (since the map $\bar{x}$ is onto) there is an m' such that $w = \bar{x}(m')$, and by the direction of (6) already proved $m' = \underline{x}(w) = m$. So $w = \bar{x}(m)$, which proves (6).

To prove the last statement in the lemma, note that $m \leq_w \underline{x}(w)$ implies by definition that $\sum_{i <_w m} x_{iw} = 0$. So $\sum_i x_{iw} = \sum_{i >_w m} x_{iw} + x_{mw} = 1$. But if $\sum_{j >_m w} x_{mj} + \sum_{i >_w m} x_{iw} + x_{mw} = 1$, this implies $\sum_{j >_m w} x_{mj} = 0$, that is, $w \geq_m \bar{x}(m)$, which completes the proof.

We are now in a position to prove Theorem 3.22.

Proof of Theorem 3.22: As noted following the statement of the theorem, what remains to be shown is that every extreme point of C is integer. So let x be an extreme point of C, and define the $|M| \times |W|$ matrices $\bar{z}$, $\underline{z}$, and z by, for each m in M and w in W:

$$\bar{z}_{mw} = 1 \text{ if } w = \bar{x}(m) \text{ and } \bar{z} = 0 \text{ otherwise,}$$

$$\underline{z}_{mw} = 1 \text{ if } w = \underline{x}(m) \text{ and } \underline{z}_{mw} = 0 \text{ otherwise,} \quad \text{and}$$

$$z_{mw} = \bar{z}_{mw} - \underline{z}_{mw}.$$

Note that Lemmas 3.23 and 3.24 imply that $\bar{z}_{mw} = 1$ if and only if $m = \underline{x}(w)$, and $\underline{z}_{mw} = 1$ if and only if $m = \bar{x}(w)$.

So for all m in M, $\sum_j \bar{z}_{mj} = \sum_j \underline{z}_{mj} = 1$, and for all w in W, $\sum_i \bar{z}_{iw} = \sum_i \underline{z}_{iw} = 1$, and so from the definition of z we have

(10) $$\sum_i z_{iw} = \sum_j z_{mj} = 0 \text{ for all } m \text{ in } M \text{ and } w \text{ in } W.$$

(11) Also note that if $x_{mw} = 0$, then $z_{mw} = 0$ (since $\bar{z}_{mw} = \underline{z}_{mw} = 0$).

The plan of the proof is now to show that for sufficiently small positive δ, $x + \delta z$ and $x - \delta z$ satisfy (1)–(4). Since $x = [(1/2)x + \delta z + (1/2)x - \delta z]$ for all δ, the fact that x is an extreme point implies that $x + \delta z = x - \delta z = x$, that is, $z = 0$. But this implies $\bar{z} = \underline{z}$, that is, $\bar{x}(m) = \underline{x}(m)$ for all m in M, and so x is integer.

Conditions (10) and (11) show that the matrices $x \pm \delta z$ satisfy constraints (1), (2), and (4) for sufficiently small δ. To show that constraint (3) is also satisfied, we need to establish, for all m in M and w in W, that

(12) if $$\sum_{j >_m w} x_{mj} + \sum_{i >_w m} x_{iw} + x_{mw} = 1,$$

then $$\sum_{j >_m w} z_{mj} + \sum_{i >_w m} z_{iw} + z_{mw} = 0.$$

We will suppose that for some m and w the left-hand sum of (12) equals 1, and prove that the right-hand sum equals 0 by considering three cases. First, suppose $w \geq_m \bar{x}(m)$. Then

$$\sum_{j >_m w} \bar{z}_{mj} = 0,$$

and, since $w \geq_m \bar{x}(m) \geq_m \underline{x}(m)$,

$$\sum_{j >_m w} \underline{z}_{mj} = 0.$$

But by Lemma 3.23, $w \geq_m \bar{x}(m)$ implies $m \leq_w \underline{x}(w) \leq_w \bar{x}(w)$ so [since $\bar{z}_{iw} = 1$ if and only if $i = \underline{x}(w)$, and $\underline{z}_{iw} = 1$ if and only if $i = \bar{x}(w)$],

$$\sum_{i>_w m} \bar{z}_{iw} + \bar{z}_{mw} = \bar{z}_{\underline{x}(w), w} = 1, \quad \text{and}$$

$$\sum_{i>_w m} \underline{z}_{iw} + \underline{z}_{mw} = \underline{z}_{\bar{x}(w), w} = 1.$$

Since $z = \bar{z} - \underline{z}$, that the right-hand sum of (12) equals 0 follows immediately from the preceding four equations. This establishes (12) for the case $w \geq_m \bar{x}(m)$. (Note that for this case we have not needed the assumption that the left-hand sum of (12) equals 1.)

The case $w \leq_m \underline{x}(m)$ now proceeds by noting that since the left-hand sum of (12) equals 1, Lemma 3.24 implies that $m \geq_w \bar{x}(w)$, and (12) now follows from arguments just like those just used, but exchanging the roles of m and w.

The final case is $\underline{x}(m) <_m w <_m \bar{x}(m)$. Then

$$\sum_{j>_m w} \bar{z}_{mj} = 1 \quad \text{and} \quad \sum_{j>_m w} \underline{z}_{mj} = 0.$$

Since the left-hand sum of (12) equals 1, Lemma 3.23 implies $\underline{x}(w) <_w m$. Since $\underline{x}(m) <_m w$, Lemma 3.24 implies that $m <_w \bar{x}(w)$. So [since $\bar{z}_{iw} = 1$ if and only if $i = \underline{x}(w)$, and $\underline{z}_{iw} = 1$ if and only if $i = \bar{x}(w)$],

$$\sum_{i>_w m} \bar{z}_{iw} = 0 \quad \text{and} \quad \sum_{i>_w m} \underline{z}_{iw} = 1.$$

Since $\bar{z}_{mw} = \underline{z}_{mw} = 0$, the preceding four equations and the definition $z = \bar{z} - \underline{z}$ implies that the right-hand sum of (12) equals 0. This completes the proof.

Note that a consequence of Theorem 3.22 is that the stable matching that maximizes a linear objective function can be computed by linear programming. (E.g., it is a linear programming problem to find the stable matching x that maximizes the sum over all m and w of terms of the form $c_{mw} x_{mw}$, where c_{mw} is some function of m's ranking of w and w's ranking of m.) This might seem to be an invitation to certain kinds of central planning. However, as we shall see, this prospect is largely illusory. The problem is that the information the planner requires, which is the preference ordering of each agent, is typically available only by asking the agents what their preferences are. And as we shall see in the following chapters, theory suggests that, in a planning problem of this kind, agents will have incentives not to reveal to the planner precisely what he or she wants to know, and empirical studies confirm that agents act on such incentives. Indeed, one of the principal results of modern theoretical and empirical research on matching is that the incentives that any matching procedure gives to the agents impose serious constraints on what kinds of plans can be implemented. As we shall see in the following chapters, much of the importance of stability itself comes from these incentive constraints.

3.3 Guide to the literature

Theorem 3.4 is due to Demange, Gale, and Sotomayor (1987). Some closely related results are found in Demange (1987). The statement of Lemma 3.5, the blocking lemma, is due to J. S. Hwang (n.d.), and was proved with the two proofs presented here in Gale and Sotomayor (1985a). Theorem 3.8 was presented in Knuth (1976). Blair's observation that the cores of marriage games exhaust the set of distributive lattices (Theorem 3.9) is contained in his 1984 paper, and further refinement of that result is presented by Gusfield, Irving, Leather, and Saks (1987). (A good general reference on the theory of lattices is Birkhoff 1973).

The core of a "housing" market that involves the exchange of indivisible goods, and which appears to fall somewhere between the roommate problem and the marriage problem, is explored in Shapley and Scarf (1974), and Roth and Postlewaite (1977). In that model, each individual owns one unit of an indivisible good, and has preferences over all the goods. An outcome is a matching of people with goods. Shapley and Scarf showed that the core of this market is always nonempty, and Roth and Postlewaite showed that when preferences are strict, the core defined by weak domination (see Section 5.7) consists of a single point that coincides with the unique competitive allocation. Much of the game-theoretic structure of this model comes from the fact that individuals come to the market owning a good, which they may keep if nothing preferable is offered. A class of problems involving the assignment of people to "positions" that are not initially owned by anyone is considered from various perspectives by Gale (1968), Gardenfors (1973), Hylland and Zeckhauser (1979), Proll (1972), Wilson (1977), and Francis and Fleming (1985).

So far we have referred only to the most closely related literature. However, as noted in Section 3.1, the material in this chapter is related to the wider game-theoretic and economic literature through the fact that the set of stable matchings equals the core of the market game. The core is one of the most important analytic tools of cooperative game theory. The idea of the core goes back at least to Edgeworth's (1881) "contract curve." The idea was briefly discussed by von Neumann and Morgenstern (1944) in the context of a general solution for games, and studied as a distinct solution concept for games by Gillies (1953a, b) and Shapley (1953a). Its importance in economics stems in part from the fact that, in markets, every allocation corresponding to a competitive equilibrium is contained in the core. The core can be thought of as generalizing the set of competitive allocations. In general markets, the core contains other allocations as well, whereas in models of perfectly competitive economies, consisting of infinitely many small agents, the core coincides with the set of competitive allocations (see, e.g., Shubik 1959, Debreu and Scarf 1963,

Aumann 1964). (Kaneko and Wooders 1986, 1985 consider some models of two-sided matching with a continuum of players, and show that the core is nonempty.) An introductory treatment of some of these general game theoretic issues can be found in Shubik (1982). Some elementary classroom exercises related to the marriage problem are presented in Brissenden (1974).

In recent years, following the work of Peleg (see Peleg 1988 for a review), a number of axiomatizations of the core for different classes of games have appeared. Toda (1988) and Sasaki (1988) axiomatize the set of stable outcomes for the marriage problem in this way, showing that it is the unique set of outcomes that is "consistent" in certain ways having to do with comparisons of markets of different sizes. One of the key results has to do with the addition of agents to the market. Toda (1988) shows that for any stable matching μ, it is possible to add to the market agents with preferences such that the larger market has a unique stable matching, which coincides with μ on the original set of agents.

The algorithm of Section 3.2.1 is from Irving and Leather (1986). The results and proofs given here are modifications of those that they presented. They in turn follow the algorithm for computing all stable matchings presented by McVitie and Wilson (1971). The results of Section 3.2.2 are from the same paper. Theorem 3.20 and the associated algorithm are due to Gusfield (1987).

Some other computational questions, and different approaches to implementing deferred acceptance procedures, are considered by Knuth (1976), and by McVitie and Wilson (1970b), Wilson (1972, 1977), Itoga (1978, 1981, 1983), Allison (1983), Hull (1984), Tseng and Lee (1984), Kapur and Krishnamoorthy (1985), Quinn (1985), and Gusfield and Irving (1989). Itoga (1981) discusses the problem of recomputing a stable matching when new men or women are added to a marriage problem, starting from a stable matching of the original problem. Itoga (1983) and Quinn (1985) consider how the computational burden is eased if some instabilities may be permitted. Irving (1985) and Gusfield and Irving (1989) consider computational problems associated with the roommates problem. Although a good deal of attention has been given to the computational questions associated with worst case performance of algorithms to compute stable matchings (see Garey and Johnson 1979 for an introduction to the style of analysis), as yet relatively little seems to be known about average performance. Some innovative work using a network formulation is contained in Feder (1989) and Subramanian (1989).

Theorems 3.21 and 3.22 in Section 3.2.4 are from Vande Vate (1988). His work builds on that of Irving, Leather, and Gusfield (1987), who were concerned with identifying "optimal" stable matchings along the lines discussed in the last paragraph of Section 3.2.4. The proofs given here, using Lemmas 3.23 and 3.24 are from Rothblum (1989). Rothblum notes that these two lemmas can be used to prove the decomposition lemma (in the form of Corollary 2.21): If x and y are stable matchings, the trick is to apply the two lemmas to the fractional matching $z = (1/2)x + (1/2)y$. Rothblum also observes that the results of the section do not depend on the simplifying assumptions that there are equal numbers of men and women who are all mutually acceptable. In the more general formulation, constraints (1) and (2) are inequalities, $x_{mw} = 0$ for any pair (m, w) that is not mutually acceptable, and constraint (3) applies only to mutually acceptable m and w. Roth, Rothblum, and Vande Vate (1990) explore further consequences of the linear structure of stable matching, and observe that a number of results we have obtained by other means can be deduced as consequences of the duality and complementary slackness theorems of linear programming.

CHAPTER 4

Strategic questions

We turn now to a different class of questions, which will require different kinds of models and theories. In the previous sections we investigated the marriage market by exploring the kind of matchings we might expect to observe. Now we will investigate how we should expect individual agents to behave. Specifically, we will want to know to what extent it is wise for men and women to be frank about their preferences for possible mates, and how we can expect them to act in the process of courtship and marriage. Lest the analysis seem a little cold-blooded at times, it is good to remember that our primary interest here is in a simple model of labor markets, and that the phenomena we will be studying can perhaps better be thought of as the courtship that takes place between potential employers and employees. But for simplicity we consider these questions first in the context of the marriage problem, and defer to later chapters more realistic models in which firms may employ more than one worker.

To address these questions of individual behavior, we need to model the decisions that individuals may be called upon to make in the course of a marriage market. We have so far considered only the general rules of the game, which state that for a marriage to take place between some man m and woman w, it is both necessary and sufficient that the two of them agree. These rules are reflected in our definition of a stable matching. But these general rules might be implemented by many different particular procedures. These particular procedures, which constitute the detailed rules of the game, will determine what specific decisions each agent faces. They will determine how an agent goes about making his preferences known, and the order in which decisions are taken. They will, in short, be a description of the mechanics of the market.

To get an idea of the kind of question to be explored here, consider the deferred acceptance algorithm used in Section 2.3 to prove the existence of stable matchings (and also of M- and W-optimal stable matchings). That algorithm was described in terms of certain actions of the agents –

proposals by the men, and acceptances and rejections by the women. The purpose of our earlier discussion of the algorithm was simply to demonstrate that stable matchings always exist, and so we did not have to consider whether it was sensible for men and women to act as described in the algorithm, in a game with appropriate rules for making and accepting proposals. Let us now turn to this question.

Consider the following rules.

1. Actions in the market are organized in stages. Each stage is divided into two periods. Within each period, each man and woman must make decisions without knowing the decisions of other men and women in that period.

2. At the first period of the first stage, each man may make at most one proposal to any woman he chooses. (He is also free to make no proposals.) Proposals can only be made by men.

3. In the second period of the first stage, each woman who has received any proposals is free to reject any or all of them immediately. A woman may also keep at most one man "engaged" by not rejecting his proposal.

4. In the first period of any stage, any man who was rejected in the preceding stage may make at most one proposal to any woman he has not previously proposed to (and been rejected by). In the second period, each woman may reject any or all of these proposals, including that of any man who has proposed in an earlier stage and been kept engaged. A woman may keep at most one man engaged by not rejecting his proposal.

5. If, at the beginning of any stage, no man makes a proposal, then the market ends, and each man is matched to the woman he is engaged to. Men who are not engaged to any woman, and women who are not engaged to any man, remain single.

A complete description of the rules of the game also includes a description of what information each agent possesses at different points in the game. For now let us suppose that at each stage each man who has made a proposal learns only whether he has been rejected or kept engaged, and each woman learns only which men have proposed to her.

Note that the rules of the game are stated without reference to the preferences of the agents. This corresponds to our usual notion of rules – they apply to any agents involved in the market, regardless of what their preferences might be. Our earlier discussion of the deferred acceptance algorithm, in contrast, also included a description of a certain way in which men and women might behave in each period as a function of their preferences. In the algorithm, each man proposed at each period to the woman he liked best of those remaining, and each woman rejected at each period all but the man she liked best among the acceptable men who had so far proposed to her. This behavior, which we will call "straightforward behavior," was shown to lead to a stable matching for all possible preferences.

The question we want to ask now is:

> Is it always in each agent's best interest, under these rules, to behave in this straightforward manner?

If the answer to this question is No, there will be at least some occasions in which it would be profitable for a man or woman to behave unstraightforwardly. It might be that some man, considering the matching that will result if he behaves straightforwardly, could be matched to a woman he prefers if at some step of the market he proposed (instead) to some woman other than the one he likes best among those remaining. Or it might be that some woman would find it in her interest to keep engaged, at some step of the market, some man other than the one she likes best among her proposers up to that stage. We will refer to this kind of unstraightforward behavior as "strategic behavior."

Another way to approach the same issue is to consider a marriage market organized by a matchmaker (or by a computer dating service – recall the description of the National Resident Matching Program in Section 1.1, to which we shall return in Chapter 5). Since the matchmaker does not know the preferences of the men and women, the matchmaker intends to get this information by asking each eligible man and woman to rank in order of preference each prospective mate. Suppose that the matchmaker will then employ the deferred acceptance algorithm (with men proposing), using the stated preferences, to produce the set of marriages. In this situation we can formulate our question as follows.

> Is it always in each agent's best interest to state his or her *true* preferences to the matchmaker?

If the answer to this question is No, there will be situations in which some man or woman could be matched with a more preferred mate if he or she behaved "strategically" by stating some ranking of prospective mates that differs from his or her true preferences.

These two questions are closely related. If it were always in an agent's best interests to behave in a straightforward way in the "decentralized market" (i.e., without a matchmaker) under the rules described, then it would also be in his (or her) best interest to state his true preferences to a matchmaker who uses the deferred acceptance algorithm, since in this case the matchmaker will assign him to the same mate he would have had by straightforward behavior in the decentralized market.

It turns out that we can answer these questions quickly by looking at an appropriate example. In Example 2.9, which we used before to illustrate the deferred acceptance algorithm (with men proposing), one of the women can do better if she misrepresents her preferences to the matchmaker (or, equivalently, if she doesn't play straightforwardly in the decentralized game).

Example 4.1: An example in which there is an incentive to misrepresent
Recall that in Example 2.9, the M-optimal stable matching is

$$\mu_M = \frac{w_1 \ w_2 \ w_3 \ w_4 \ (m_5)}{m_1 \ m_2 \ m_3 \ m_4 \ m_5}.$$

This matching matches woman w_1 to man m_1, her third choice.

Consider now the preferences P' in which all agents except w_1 state their preferences as in Example 2.9, but w_1 misrepresents her preferences by stating

$$P'(w_1) = m_2, m_3, m_4, m_5, m_1$$

instead of her true preference

$$P(w_1) = m_2, m_3, m_1, m_4, m_5.$$

Then the resulting matching μ'_M is

$$\mu'_M = \frac{w_1 \ w_2 \ w_3 \ w_4 \ (m_5)}{m_3 \ m_1 \ m_2 \ m_4 \ m_5}.$$

Note that w_1 has profited by misrepresenting her preferences. If she had stated her true preferences $P(w_1)$, she would have been matched to her third choice, m_1. Instead, she has ended up matched to m_3, her true second choice. So we have demonstrated that, at least sometimes, "honesty" may not be the best policy. This raises a number of new questions.

> Might there be some other way of organizing the market so that honesty *would* always be the best policy?

> In the deferred acceptance procedure, what exactly are the incentives to behave unstraightforwardly? Which agents have them, and when?

> How does unstraightforward behavior on the part of one agent affect the welfare of the other agents?

Finally, and perhaps most fundamentally, we shall have to consider whether the answers to these questions will require us to reconsider the status of the theory of stable matchings derived in the previous chapters. If agents are misrepresenting their preferences, then perhaps procedures that appear to yield stable outcomes will turn out not in fact to do so. That is, the preferences with which we wish to define stability are the *true* preferences of the agents. But if agents systematically misrepresent their preferences to, say, a matchmaker who uses the deferred acceptance algorithm, then the matching produced will be the *M*-optimal stable matching with respect to the *stated* preferences. We will need to explore circumstances under which this matching will be stable with respect to the *true* preferences, in order to know whether we can continue to expect that "stable" matchings will occur. (Although we have a number of other issues that need to be explored first, note in passing that the matching produced in Example 4.1 is in fact stable with respect to the true preferences.)

We turn now to the formal consideration of these issues.

4.1   A formal strategic model

Consider the first of the questions raised by Example 4.1: Does there exist any way to design a "sensible" matching procedure in which agents can do no better than to reveal their true preferences? We will have to specify what we mean by "sensible" of course, but it is clear that some such qualification will be needed. (Consider, for example, a matching procedure that results in every man and woman remaining single regardless of what agents say or do. In this procedure, no agent can do better than to reveal his true preferences.) In order to explore this question properly, we need a more formal model of the decisions facing agents in a matching market.

It will be convenient to consider this problem from the point of view of designing an algorithm to be used by a matchmaker. That is, in this section we will consider matching markets in which each agent states to the matchmaker some preference ordering over possible mates. The matchmaker then uses these stated preferences as the input to some algorithm that produces the final matching. We have already observed informally that answering this question for a matchmaker market will also tell us the answers to similar questions posed about markets organized along decentralized lines. We will return to this point more formally in Section 4.5, when we talk about the "revelation principle."

Consider a marriage market (M, W, P) whose outcome will be determined by a matchmaker who will employ some particular algorithm to produce a matching, based on a list of preference orderings that agents will state. That is, each man m, whose preferences are $P(m)$, is faced with the problem of deciding what preference ordering $Q(m)$ to state, and each woman w with preferences $P(w)$ must state a preference ordering

$Q(w)$. The set of stated preference lists, one for each man and woman, will be denoted by $Q = \{Q(m_1), ..., Q(m_n), Q(w_1), ..., Q(w_p)\}$. The algorithm employed by the matchmaker produces a matching as a function of the stated preferences Q. That is, the matchmaker produces a matching $\mu = h(Q)$, where h is the function that describes the output of the matchmaker's algorithm, for any set Q of stated preferences.

We have just described the market in what game theorists call *strategic form*. In general, the components of the strategic form description are a set N of players, a set of feasible strategies D_i for each player i in N, an outcome function h that describes how choices of strategies by the players determine an outcome, and preferences for each player over the set of outcomes. The players in this case are the set $N = M \cup W$ of men and women. In this case, the simplest interpretation will be to take the set of strategies available to each player to be the set of all preference lists he or she might state. The *outcome* of a given set Q of stated preferences is the matching $\mu = h(Q)$, where h is the outcome function that describes how the matchmaker's algorithm turns sets of stated preferences into a matching, and true preferences of the players are given by P.

We have not treated the matchmaker as a player in the game. Instead, the matchmaker is modeled as a purely mechanical part of the environment, with no decisions to make. The "matchmaker" is simply a surrogate for whatever mechanism exists for turning decisions of the men and women into matches. We will discuss in Section 4.5 why this is a fruitful approach. Essentially what we will see there is that Theorem 4.4 (to follow) applies to any matching market, whether it is organized under decentralized rules or with some central matchmaking mechanism (as in the hospital–intern market).

Note also that the strategic form $(N = M \cup W, \{D_i\}, h, P)$ of a game is a more detailed model than the model (M, W, P) presented in Chapter 2, since it includes information about how outcomes result from specific decisions of the agents. However it is still by no means the most detailed model we could have, since the decisions of the players are represented in a very terse form, as a single decision of what strategy to choose.

4.1.1 *Dominant strategies*

Some notation will help focus on the actions of individual agents. If Q represents the choices of all the agents, and we want to focus on the decision facing one of them, say individual i, then we will sometimes write $Q = (Q_{-i}, Q(i))$, where $Q(i)$ is the choice of player i, and Q_{-i} is the set of choices of all agents other than i. In this way we can consider alternative sets $Q' = (Q_{-i}, Q'(i))$ of strategy choices, which differ from Q only in player i's choice.

To address questions such as Is honesty the best policy? we need to develop ideas about what constitutes a "best policy." Consider for a moment the problem that would face player i if the set of strategy choices Q_{-i} of all the other players were known. In this case, any choice of strategy $Q(i)$ by i would determine the outcome $h(Q_{-i}, Q(i))$. A particular strategy choice $Q^*(i)$ by player i is a *best response by i to Q_{-i}* if player i likes $h(Q_{-i}, Q^*(i))$ at least as well as any of the outcomes $h(Q_{-i}, Q(i))$ that would have resulted from any other strategy $Q(i)$ he or she could have chosen. [Note that we evaluate whether $h(Q_{-i}, Q^*(i)) \geq_i h(Q_{-i}, Q(i))$ according to the preference relation $\geq_i$ corresponding to i's *true* preferences $P(i)$.]

Of course, a strategy $Q^*(i)$ that is a best response to Q_{-i} might not be a good response at all if the other agents make a different set of strategy choices. What makes multiperson decision problems so different from single person decisions is precisely that an individual's best course of action typically depends critically on what other people do. So in a problem like the one described here, in which each agent must make a decision without knowing what the other agents have done, it may be that there doesn't exist anything that can be called a "best policy" for agent i. But if agent i should happen to have a strategy that was a best response to *any* set of strategy choices that the other players might make, then it would clearly be his or her best policy to choose this strategy. When such a strategy exists, it is called a *dominant strategy*. Formally,

Definition 4.2. *A dominant strategy for an agent i is a strategy $Q^*(i)$ that is a best response to all possible sets of strategy choices Q_{-i} by the other agents.*

Thus, if man m has some dominant strategy $Q^*(m)$, then by choosing it he can always get the best outcome obtainable for him given the strategies chosen by the other agents, and so he has no incentive to play any other strategy.

In discussing any algorithm that might be employed by the matchmaker, we will only need to be concerned with the matching produced for any set of stated preferences, since the agents are concerned only with the outcome. That is, the internal workings of the algorithm are of concern here only as they affect how the data of the problem are translated into a final matching. Thus we need only consider the function that, for each input, describes the matching that the algorithm produces. Such a function will be called a *matching mechanism*. Formally, a matching mechanism is a function h whose range is the set of all possible inputs (M, W, Q), and whose "output" $h(Q)$ is a matching between M and W. If $h(Q)$ is always stable with respect to Q, it will be called a *stable* matching mechanism. If $h(Q)$ is always Pareto optimal with respect to the set of agents (i.e., if no other matching makes all men and women better off according to the preferences Q), then it will be called a *Pareto optimal* matching mechanism. Note that stable and Pareto optimal matching mechanisms are defined with respect to the *stated* preferences Q, which are the mechanism's input.

If a particular matching mechanism h is adopted for use with a set of players M and W who have preferences P, it serves as the outcome function in the strategic game $(N = M \cup W, \{D_i\}, h, P)$ we have just described. We say that game is *induced* by the mechanism h. Of course the agents evaluate matchings in terms of their true preferences P. These true preferences determine their behavior. A matching mechanism will be called *strategy proof* if it makes it a dominant strategy for each player to state his or her true preferences in the strategic game it induces. Games in which players must state their preferences are sometimes called *revelation games,* and the mechanism h is called a *revelation mechanism.*

4.2   The impossibility of a "strategy proof" stable mechanism

Here we are going to show that honesty is not the best policy (at least not all the time for every agent) when the game is built around a mechanism that yields stable outcomes. But first, to see the role of stability, consider a game induced by a Pareto optimal matching mechanism in which honesty *is* the best policy (or at least as good as any other). The example resembles the procedures used by professional sports teams in the United States to draft college players (although the conclusions of the example depend on the fact that we are considering the case of the marriage market, in which matching is one-to-one.)

Example 4.3: A strategy proof, Pareto optimal matching mechanism (Roth)
Let the men be placed in some order, $m_1, ..., m_n$. Consider the mechanism that for any stated preferences Q yields the matching $\mu = h(Q)$ that matches m_1 to his stated first choice, m_2 to his stated first choice of possible mates remaining after $\mu(m_1)$ has been removed from the market, and any m_k to his stated first choice after $\mu(m_1)$ through $\mu(m_{k-1})$. It is clearly a dominant strategy for each man to state his true preferences, since each man is married to whomever he indicates is his first choice among those remaining when his turn comes. It is also (degenerately) a

dominant strategy for each woman to state her true preferences, since the preferences stated by the women have no influence. The mechanism h is Pareto optimal, since at any other matching some man would do no better. However h is not a stable matching mechanism, since it might happen, for example, that woman $w = \mu(m_1)$, who is the (draft) choice of man m_1, would prefer to be matched with someone else, who would also prefer to be matched to her. That is, h is not a stable matching mechanism because there are some sets of preferences for which it will produce unstable outcomes. [In football drafts, the team with the worst record gets the first draft choice, so presumably the worst team gets the best player. If (as seems to be the case) most players rank the worst team near the bottom of their preferences, and most teams rank the best player near the top of their preferences, the resulting matching is unstable in the sense defined here. But since the rules of the football market do not give the players the right to negotiate with any team they choose, no blocking pairs are allowed to form.]

An almost equivalent procedure is the one used at the United States Naval Academy, by which graduating students obtain their first posts as Naval officers. The following description is taken from the *New York Times* (January 30, 1986, p.8).

Midshipmen who will graduate from the Naval Academy in June decided this week whether they wanted to be aviators or nuclear submariners, destroyermen or engineers, marines or oceanographers.... From late Thursday afternoon through the wee hours of Friday morning, the first classmen, or seniors, lined up according to their standing in the class, walked up to a long table lined with officers from each specialty, and made their choices on a first-come, first-served basis.... Emotions sometimes ran high in the selection. There was elation for those who got their first choices, tension in the middle of the class for midshipmen who had to make spot decisions depending on what was still open, and disappointment for some of those at the bottom.

Note, however, that unlike the football draft, it is the *best* student (rather than the worst team) who gets to make the first choice. If the representative of every Naval specialty were to prefer the midshipmen in the order of their class standing, the procedure would in that case yield a stable matching. (Of course, although it is not unreasonable to expect that preferences for midshipmen often correspond closely to their class standing, one would need to investigate carefully the specific market in question before trying to draw conclusions on this basis.) In any event, the procedure is still not a stable matching mechanism, since there are some sets of preferences for which it would yield an unstable matching.

Two brief observations on modeling are in order before we turn our attention once again to stable matching mechanisms. First, note that the situation at the Naval Academy differs from Example 4.3 in that it is not a marriage problem, since each Naval specialty may accept more than one midshipman. However it is easy to see that this does not alter our conclusion that each midshipman has a dominant strategy, since he or she receives the position chosen from the positions still available when he or she reaches the head of the line. More importantly, notice that the Naval Academy procedure is not a matchmaker procedure in which each midshipman states a full list of preferences to some intermediary. Instead, each midshipman makes a choice only when his or her turn comes, and at least the *New York Times* journalist believes that some of the midshipmen made "spot decisions" at the last moment. Here, too, it should be reasonably clear that a full set of contingent decisions of the form, What specialty should I choose from those that remain when I get to the head of the line? is equivalent to deciding what preference ordering to state to a matchmaker.

4.2.1 *Stable matching mechanisms*

The situation for stable matching mechanisms is quite different. We will now see that the observation in Example 4.1 that women can sometimes profit by misrepresenting their preferences is not peculiar to the deferred acceptance procedure. The fact that the deferred acceptance procedure can sometimes be profitably "manipulated" by an agent's choice of what preferences to state is a consequence of the fact that it is a stable matching mechanism. To put it another way, it is impossible to design a mechanism that produces stable outcomes (in terms of the stated preferences), and that also makes it a dominant strategy for every agent to state his or her true preferences.

Theorem 4.4: Impossibility theorem (Roth). *No stable matching mechanism exists for which stating the true preferences is a dominant strategy for every agent.*

Since a matching mechanism is a function that produces a matching for *any* marriage market, to prove the theorem it is sufficient to demonstrate some particular marriage market such that for any possible stable matching mechanism, truth telling is not a dominant strategy for all agents. This is done in the following proof, which considers a marriage market with two men and two women, and shows that every stable matching mechanism makes it possible for some man or woman to profit by misstating his or her preferences. (The result therefore follows for any larger marriage market, since nothing changes if these four agents are embedded in a larger market, without changing their preferences.)

Proof: Consider a market with two men and two women with preferences P given by $P(m_1) = w_1, w_2$; $P(m_2) = w_2, w_1$; $P(w_1) = m_2, m_1$; $P(w_2) = m_1, m_2$. Then there are two stable matchings, μ and ν, given by $\mu(m_i) = w_i$ for $i \in \{1, 2\}$, and $\nu(m_i) = w_j$ for $i, j \in \{1, 2\}$, $j \neq i$. So any stable mechanism must choose one of μ or ν when preferences P are stated. Suppose the mechanism chooses μ. Observe, for example, that if w_2 changes her stated preference from $P(w_2)$ to $Q(w_2) = m_1$ while everyone else states their true preferences, then ν is the only stable matching with respect to the stated preferences $P' = (P(m_1), P(m_2), P(w_1), Q(w_2))$, and so any stable mechanism must select ν when the stated preferences are P'. So it is not a dominant strategy for all agents to state their true preferences, since w_2 does better to state $Q(w_2)$. Similarly, if the mechanism chooses ν when the preferences P are stated, then m_2 can profitably misstate his preferences.

An immediate corollary of the impossibility theorem can be stated as follows.

Corollary 4.5.   *No stable matching mechanism exists for which stating the true preferences is always a best response for every agent when all other agents state their true preferences.*

Since a matching mechanism is a function defined over marriage problems with all possible preferences, if stating the truth were always an agent's best response when other agents state the truth, then stating the truth would be a dominant strategy for the mechanism, in contradiction to the impossibility theorem.

Since we have defined a matching mechanism as a procedure that can be applied to any marriage market (i.e., as a function defined for all marriage markets), the impossibility theorem says we cannot find a stable mechanism that won't *sometimes* give some agent an incentive to misstate his or her preferences. But we might hope to find a stable matching mechanism that only seldom gave agents such incentives, in which case the problem of incentives might not be very important. The following result, which can be thought of as a generalization of the proof of the impossibility theorem, and which strengthens it, states that no such mechanism can be found. Instead, that at least one agent will have an incentive to behave strategically seems to be the usual case.

Theorem 4.6.   *When any stable mechanism is applied to a marriage market in which preferences are strict and there is more than one stable matching, then at least one agent can profitably misrepresent his or her* *preferences, assuming the others tell the truth. (This agent can misrepresent in such a way as to be matched to his or her most preferred achievable mate under the true preferences at every stable matching under the misstated preferences.)*

Proof: By hypothesis $\mu_M \neq \mu_W$. Suppose that when all agents state their true preferences, the mechanism selects a stable matching $\mu \neq \mu_W$. Let w be any woman such that $\mu_W(w) >_w \mu(w)$. (So w is not single at μ_W.) Now let w misrepresent her preferences by removing from her stated preference list of acceptable men all men who rank below $\mu_W(w)$.

Clearly the matching μ_W will still be stable under these preferences (there are now fewer possible blocking pairs). Letting μ' be the stable matching selected by the mechanism for these new preferences, it follows from Theorem 2.22 that w is not single at μ' and hence she is matched with someone she likes at least as well as $\mu_W(w)$, since all other men have been removed from her list of acceptable men. In fact, since μ' is also stable under the original preferences, it follows that $\mu'(w) = \mu_W(w)$. But w prefers $\mu_W(w)$ to $\mu(w)$ so she prefers any stable μ' under the new preferences to μ. If the mechanism originally selects the matching μ_W, then the symmetric argument can be made for any man m who strictly prefers μ_M. This completes the proof.

When preferences need not be strict, the conclusions of the theorem need not hold – the easiest counterexample is one in which all women are indifferent between the different stable matchings.

4.3 The incentives facing the men when an *M*-optimal stable matching mechanism is employed

The impossibility theorem states that any stable matching mechanism will sometimes give some agents an incentive to misrepresent their preferences. Example 4.1 and the proof of Theorem 4.6 showed that under the deferred acceptance procedure with men proposing, some women may have such an incentive. We shall return to the women in the next section, after considering first the incentives facing the men under this procedure. Since the procedure treats men and women differently, it is reasonable to conjecture that the strategic problems facing men and women are not the same.

The important feature of the deferred acceptance procedure with men proposing is that when preferences are strict, it produces the unique *M*-optimal stable matching for any stated preferences. When preferences are strict, we can therefore speak of *the* M-optimal stable matching mechanism. When preferences are not strict so that no M-optimal stable matching may exist, we need to specify more carefully what algorithm we mean. As we remarked when using the deferred acceptance procedure to prove that stable outcomes always exist, when the algorithm is supplemented with a tie-breaking procedure, it produces a stable matching even when not all preferences are strict.

The theorems in this section hold not only for the case of strict preferences, but also for the general case. For ease of exposition, however, we first consider the case of strict preferences, and so we can refer to *the* M-optimal stable matching mechanism. The assumption of strict preferences will be relaxed in stating and proving the last theorem of this section, Theorem 4.11, which will in turn provide the proof for the other theorems in this section. (The original proofs for these other theorems were rather involved.) Our first result concerns the incentives facing individual men.

Theorem 4.7 (Dubins and Freedman; Roth). *The mechanism that yields the M-optimal stable matching (in terms of the stated preferences) makes it a dominant strategy for each man to state his true preferences. (Similarly, the mechanism that yields the W-optimal stable matching makes it a dominant strategy for every woman to state her true preferences.)*

Throughout this section, $P = \{P(m_1), \ldots, P(w_m)\}$ will denote the *true* preferences of the agents, and we will write $a >_x b$ to mean that agent x prefers a to b under $P(x)$. The M-optimal stable outcome with respect to the true preferences P will be denoted μ_M. Similarly, μ'_M and μ''_M will denote the M-optimal stable matching with respect to stated preferences P' and P'', respectively.

Although we will prove Theorem 4.7 only indirectly following Theorem 4.11, two of the lemmas from the original proof are of independent interest, and we present these here.

There are of course many ways in which a man m might state a preference ordering $P'(m)$ different from $P(m)$, but the first lemma shows that, in considering man m's incentives to misstate his preferences, we can confine our attention to certain kinds of simple misrepresentations. Suppose by stating some preference list $P'(m)$, man m can change his mate from $\mu_M(m)$ to $\mu'_M(m)$. Then he can get the same result – that is, he can be matched to $\mu'_M(m)$ – by stating a preference list $P''(m)$ in which $\mu'_M(m)$ is his first choice. So, if there is any way for m to be matched to $\mu'_M(m)$ by stating some appropriate preference list, then there is a simple way – he can just list her as his first choice.

Lemma 4.8 (Roth). *Let m be in M. Let μ'_M and μ''_M be the corresponding M-optimal stable matchings for (M, W, P') and (M, W, P''), where $P''(x) = P'(x)$ for all agents x other than m, and $\mu'_M(m)$ is the first choice for m in $P''(m)$. Then $\mu''_M(m) = \mu'_M(m)$.*

Proof: Clearly the matching μ'_M is stable under the preference profile P''. Since $\mu'_M(m)$ is the first choice of $P''(m)$, $\mu''_M(m) = \mu'_M(m)$, from the optimality of μ''_M.

Lemma 4.8 shows that to prove Theorem 4.7, it is sufficient to prove that no *simple* manipulation, $P'(m)$, in which m lists $\mu'_M(m)$ first, can be *successful;* that is, we cannot have $\mu'_M(m) >_m \mu_M(m)$. The following Lemma shows that if a simple misrepresentation by m leaves *m at least as well off* as at μ_M, then no man will suffer; that is, every man likes the matching μ'_M resulting from the misrepresentation at least as well as the matching μ_M. This illustrates another way in which the men have common rather than conflicting interests.

Lemma 4.9 (Roth). *Let m be in M. Let μ'_M be the M-optimal stable matching for (M, W, P'). If $P'(x) = P(x)$ for all x other than m and $\mu'_M(m)$ is the first choice for m in $P'(m)$, and $\mu'_M(m) \geq_m \mu_M(m)$ under $P(m)$, then for each m_j in M we have $\mu'_M(m_j) \geq_{m_j} \mu_M(m_j)$.*

Since the deferred acceptance algorithm with men proposing yields the M-optimal stable matching, we can phrase the proof in terms of that algorithm.

Proof: Let $M^* = \{m_j; \mu_M(m_j) >_{m_j} \mu'_M(m_j)\}$. Suppose $M^* \neq \phi$. All m_j in M^* are matched under μ_M. Since every agent other than m states the same preferences under P and P' and m is not in M^*, it must be that all m_j in M^* are rejected by their mates under μ_M at some step of the algorithm for the market (M, W, P'). Let s be the *first* step of the algorithm for (M, W, P') at which some m_j in M^* is rejected by $w = \mu_M(m_j)$. Since m_j and w are mutually acceptable, this implies that w must have received a proposal in step s of the algorithm for (M, W, P') from some m_k who did not propose to her under P and whom she likes more than m_j. The fact that m_k did not propose to w under P means that $\mu_M(m_k) >_{m_k} w$. Then m_k is in M^* for if not we would have the contradiction

$$w \geq_{m_k} \mu'_M(m_k) \geq_{m_k} \mu_M(m_k) >_{m_k} w.$$

So $m_k \neq m$ and $P(m_k) = P'(m_k)$ and m_k must have been rejected by $\mu_M(m_k)$ under P' prior to step s, which contradicts the choice of s as the first such period. Consequently, $M^* = \phi$ and $\mu'_M(m_j) \geq_{m_j} \mu_M(m_j)$ for all m_j in M.

Theorem 4.7 states that, as far as each man is concerned, there is nothing he can say to a matchmaker using an M-optimal mechanism that will leave him better off than stating his true preferences. So, for men, there is a clear sense in which honesty is the best policy under such a mechanism. The next theorem turns to a different and more complex question about the incentives facing the men. Even though no individual man can profit from misrepresenting his preferences, is there any way that some *coalition* of men can obtain some matching that they all prefer to μ_M? By a coalition we mean any subset $\bar{M}$ of the set M of all men. The following theorem says that the answer to this question is also negative: Any coalition of men who each misrepresents his preferences to obtain a matching different from μ_M must contain at least one man who likes μ_M at least as well as the new matching.

Theorem 4.10 (Dubins and Freedman).   *Let P be the true preferences of the agents, and let $\bar{P}$ differ from P in that some coalition $\bar{M}$ of the men misstate their preferences. Then there is no matching μ, stable for $\bar{P}$, which is preferred to μ_M by all members of $\bar{M}$.*

The theorem implies that if the M-optimal stable mechanism is used, then no man or coalition of men can improve the outcome for all its members by misstating preferences. (Consequently this theorem implies Theorem 4.7. However we will see in Section 4.3.1 that the result for *coalitions* of men is in an important sense less robust than the result for *individual* men, since it depends critically on the assumption of the model that no side payments of any sort can be made.)

Recall the blocking lemma (Lemma 3.5), which will be needed here:

Blocking lemma.   *Let μ be any individually rational matching with respect to strict preferences P, and let M' be all men who prefer μ to μ_M. If M' is nonempty, there is a pair $\{m, w\}$ that blocks μ such that m is in $M - M'$ and w is in $\mu(M')$.*

The next theorem, which also provides the proof of the earlier theorems, asks *how far* an arbitrary stable matching mechanism can be manipulated, and by whom. We will consider coalitions that may consist of both men and women. We know that such coalitions can in general manipulate stable matching mechanisms, since the impossibility theorem tells us that even individuals can manipulate. For the case of M-optimal stable mechanisms, we saw in Example 4.1 and the proof of Theorem 4.6 that individual women (and therefore coalitions of women also) can manipulate. The next theorem says that no coalition can manipulate (by misstating preferences) so successfully that every member of the coalition prefers the new outcome to *every* stable outcome (with respect to the true preferences). This result has Theorems 4.7 and 4.10 as immediate consequences. We state and prove it for general (not necessarily strict) preferences.

Theorem 4.11: Limits on successful manipulation (Demange, Gale, and Sotomayor). *Let P be the true preferences (not necessarily strict) of the agents, and let $\bar{P}$ differ from P in that some coalition C of men and women misstate their preferences. Then there is no matching μ, stable for $\bar{P}$, which is preferred to every stable matching under the true preferences P by all members of C.*

To see that Theorem 4.11 proves Theorems 4.7 and 4.10, consider the special case in which all the coalition members are men. Then Theorem 4.11 implies that no matter which stable matching with respect to $\bar{P}$ is chosen, at least one of the liars is not better off than he would be at the *M*-optimal matching under *P*, which is what is required. (When preferences are not strict, there may of course not be an *M*-optimal stable matching, and so we have to rephrase the theorems to avoid speaking of *the M*-optimal stable mechanism. Instead, we can consider the deferred acceptance procedure with men proposing, and with a tie-breaking procedure, as in Section 2.5.1.)

One new definition will be needed in the proof of Theorem 4.11. For an agent i with true preferences $P(i)$, the strict preference list $P'(i)$ *corresponds* to $P(i)$ if the true preference list can be obtained from $P'(i)$ without changing the order of any alternatives, simply by indicating which alternatives are tied (i.e., in our notation, if $P(i)$ can be obtained by inserting brackets in the strict preference list $P'(i)$). For any profile P of preferences, not necessarily strict, we will say P' is a corresponding profile of strict preferences if each $P'(i)$ corresponds to $P(i)$.

Proof of Theorem 4.11: Suppose, by way of contradiction, that some nonempty subset $\bar{M} \cup \bar{W}$ of men and women misstate their preferences and are strictly better off under some μ, stable under $\bar{P}$ than under any stable matching under *P*. If μ is not individually rational, then someone, say a man, is matched under μ with a woman not on his true list of acceptable women, so he is surely a liar and is in $\bar{M}$, which is a contradiction. Assume now μ is individually rational. Clearly μ is not stable under

P. Now construct any corresponding profile *P'*, with strict preferences, so that, if any agent *x* is indifferent under *P* between $\mu(x)$ and some other alternative, then under *P'* *x* prefers $\mu(x)$. Then (m, w) blocks μ under *P'* only if (m, w) blocks μ under *P*. Since every stable matching under *P'* is also stable under *P*,

$$\mu(m) \underset{m}{>} \mu_M(m) \text{ for every } m \text{ in } \bar{M}, \text{ and}$$

$$\mu(w) \underset{w}{>} \mu_W(w) \text{ for every } w \text{ in } \bar{W}, \tag{1}$$

where μ_M and μ_W are, respectively, the *M*- and *W*-optimal stable matchings for (M, W, P'). If $\bar{M}$ is not empty, we can apply the blocking lemma to the market (M, W, P'), since by (1) $\bar{M}$ is a subset of *M'*; thus there is a pair $\{m, w\}$ that blocks μ under *P'* and so under *P* such that $\mu_M(m) \succeq_m \mu(m)$ and $\mu_M(w) \succeq_w \mu(w)$. Clearly *m* and *w* are not in $\bar{M} \cup \bar{W}$ and therefore are not misstating their preferences, so they will also block μ under $\bar{P}$, contradicting that μ is stable under $\bar{P}$. If $\bar{M}$ is empty, $\bar{W}$ is not empty and the symmetrical argument applies.

4.3.1   *The robustness of these conclusions*

Theorem 4.10 says that not only can no individual man do better than to state his true preferences, no coalition of men can all profit by collectively agreeing that at least some of them should misstate their preferences. In particular, the theorem states that at least one of the men who misrepresents his preferences must fail to profit from it. We remarked earlier that this conclusion about coalitions is less robust than the conclusion about individual men. To get an idea of what we mean, recall Example 2.31. In that example, the *M*-optimal stable matching μ_M was only weakly Pareto optimal from the point of view of the men; that is, there was another matching μ, not stable, that all the men liked at least as well as μ_M. In fact, except for man m_2, who was matched to woman w_3 at both μ_M and μ and was therefore indifferent between the two matchings, the other men all preferred μ.

Observe now that if man m_2 in that example were to misrepresent his preferences by listing w_3 as his first choice, then the *M*-optimal stable matching with respect to the stated preferences *P'*, in which all agents but m_2 state their true preferences, is equal to μ. That is, if m_2 misrepresents his preferences in this way under an *M*-optimal stable matching mechanism, the resulting matching is $\mu'_M = \mu$ instead of μ_M. So m_2 is able to help the other men at no cost to himself. Of course, m_2 also gets no benefit, which is what Theorem 4.10 implies, just as Lemma 4.9 implies that the men other than m_2 are not hurt by m_2's misrepresentation so long as m_2 is not.

Note, however, that if there were any way at all in which the other men could pay m_2 for his services, then it would be possible for a coalition of men to form and collectively profit from this misrepresentation. Since m_2 receives the same mate at both matchings, presumably even a very small payment would make it worth his while to become part of a coalition to change the final outcome from μ_M to μ, and since the gains to the other men in this coalition might be substantial, there would be ample motivation for such a coalition to form. Thus the implications of Theorems 4.10 and 4.11 for coalitions depend on the fact that, in the model of the marriage market that we are working with, we have assumed that no possibility exists for such "side payments" between agents. (We have also assumed that each agent is concerned only with his own mate at any matching, not with the mates of any other agents, and that the game is played only once, so that there is no possibility of a coalition forming to trade favors over time.) In the models we will explore in Part III, in which money plays a role, we will see that the implications of Theorems 4.10 and 4.11 about the lack of opportunities for coalitions to manipulate the outcome no longer hold when it is possible for agents on the same side of the market to arrange payments to each other. Indeed, much of the discussion of auction markets in Chapter 7 concerns precisely how bidders can form coalitions to influence the outcome in a manner that resembles our discussion here of Example 2.31. And in Section 4.5 we will see that coalitions of men can mutually benefit from misstating their preferences when we relax the assumption that all players know each others' preferences with certainty, since when there is even an arbitrarily small amount of uncertainty the men can trade benefits in one state of the world for those in another. In view of this, we can reinterpret Theorem 4.10 to say that in order for coalitions of men all to profit by misstating preferences, there must be some way for them to share the benefit among themselves.

However we shall see that even in these other models, the conclusion of Theorem 4.7 about individuals on the side of the market that receives its optimal stable outcome continues to hold. That is, no such individual can do better on his own than to state his true preferences.

4.4   The incentives facing the women when an M-optimal stable matching mechanism is employed

In the previous section, we saw that honesty is the best policy for the men when the M-optimal stable matching mechanism is employed. For the women, however, we saw in Example 4.1 that honesty may not be the best policy. And Theorems 4.6 and 4.7 together imply the following.

Corollary 4.12 (Gale and Sotomayor).   *When preferences are strict and the M-optimal stable mechanism is employed, there will be an incentive for some woman to misrepresent her preferences whenever more than one stable matching exists.*

To put it another way, if we were to advise all the women to state their preferences truthfully, and if they were all to follow our advice, then at least one of them could typically have done better by not following our advice. This leads us to look for other forms that advice might take, a question we consider next.

4.4.1   *Strategic equilibrium*

We will consider the following question: If honesty is not the best policy, is there any set of strategies with the property that once they are adopted by the women there will be no advantage to any one woman in changing her strategy? To put it another way, is there any advice we could give to all the women that, once it was adopted, would still be good advice for all of them to follow? A set of strategies, one for each player, that has this property is called an *equilibrium*.

Definition 4.13.   *A set of strategies* $Q = \{Q(m_1), \ldots, Q(m_n), Q(w_1), \ldots, Q(w_p)\}$ *is an **equilibrium** if, for each player i in* $M \cup W$, $Q(i)$ *is player i's best response to the strategies* Q_{-i} *of the other players.*

A set of strategies, one for each player, forms an equilibrium if no player can achieve a better payoff by changing his or her strategy, assuming the other players do not change theirs. (This notion of equilibrium, which was formulated by John Nash and is often referred to as "Nash equilibrium" is one of the most important ideas in noncooperative game theory.)

Note that the alternate statement of the impossibility theorem given in Corollary 4.5 can now be restated as follows.

Corollary 4.14.   *No stable matching mechanism exists for which it is always an equilibrium for every agent to state his or her true preferences.*

4.4.1.1 *The interpretation of equilibrium*

There are a number of ways to interpret this kind of equilibrium, which in turn have to do with the assumptions we are prepared to make about what kinds of information the agents possess. From a *prescriptive* point of view, that is, from the point of view of giving advice to the players, the attraction of equilibrium is that any other kind of advice would be self-defeating. That is, suppose our best advice consists of a set of strategies Q, one for each player, that is *not* an equilibrium. Suppose all the players decide to follow this advice. Since it is not an equilibrium, this means that at least one player is not playing his or her best response to the choices of the other players. So that player, at least, could do better than to follow our advice. If the players understand the game well, we can therefore expect that the very success of our advice among some players will cause other players not to follow it. If our advice had been equilibrium advice, however, then each player would find that following our advice to him or her was the best response when other players followed our advice to them. So equilibrium advice is the only specific advice (i.e., giving specific strategy choices to all the players) that has the chance of being good advice to all of the players simultaneously.

From a *descriptive* point of view, that is, from the point of view of trying to describe and predict how players play a game, a theory that predicts players will choose a set of strategies that is *not* an equilibrium predicts that at least one of the players will be making a mistake, in the sense that he or she could have done better by playing a different strategy. Economists are reluctant to base explanations of persistent phenomena on the hypothesis that agents are persistently mistaken about how to pursue their interests, and this reluctance naturally leads to theories that predict that some equilibrium will be chosen, since at an equilibrium, no player is making this kind of mistake. The sense in which players are not mistaken at equilibrium can perhaps be better understood by considering the *correct expectations* interpretation.

Under this interpretation, each player i forms a set of expectations, which is the set of strategies Q_{-i}, one for each of the other players, that player i expects will be chosen. He then chooses his own strategy $Q(i)$ as a best response to Q_{-i}. Let us call the resulting strategy vector Q^i. So Q^i is a vector of strategies, one for each player, which represents player i's expectation of what strategies the other players will choose, along with his or her own best response. Of course, each player j can generate a vector of strategies Q^j in this way. An equilibrium is the strategy vector Q that results when $Q^i = Q^j = Q$ for all players i and j. That is, when each player's expectations about what strategy each of the others will choose is correct, and when each player is playing his best response to these (correct) expectations, then the players are playing an equilibrium set of strategies.

Clearly the correct expectations interpretation of equilibrium makes the most sense when the players have a great deal of information about each other, and about the game they are playing, as the basis for forming correct expectations about what other players will do. There are other interpretations of equilibrium as a descriptive assumption that don't require such strong assumptions about what the players know. In general, the amount of information that players need in order to arrive at an equilibrium will depend on the equilibrating process – that is, on how the equilibrium is reached. The information demands are greatest in a game that has no history, and in which players are not able to communicate with one another, since then each player must figure out what to do from theoretical considerations alone. In contrast, evolutionary biologists are now developing game-theoretic models of strategic equilibrium in which the "strategies" are genes. (In these models, agents are not assumed to be users of information, or conscious calculators, at all. Instead, equilibrium is seen as arising from the action of differential selection, over time, among strategies that may arise randomly in a given population.) Somewhere in between are environments in which rational agents may communicate with one another before choosing their strategies (but without the possibility of reaching binding agreements). In games of this sort, equilibria are often interpreted as *self-enforcing agreements,* since once a set of equilibrium strategies has been agreed on, no player who believes the others will keep their agreement has any incentive to violate his or her part of the agreement. Equilibria can also be interpreted as being closely related to certain kinds of *social conventions,* when players are aware of the history of similar games played by similar players, and are able to use this information to form their expectations correctly. All of these interpretations potentially have a role to play in different applications of the theory of strategic equilibrium.

In order to fix ideas, we will suppose in the following section that with respect to whatever equilibrating process may be going on in the market, the players are sufficiently well informed so that they can be expected to reach some equilibrium – that is, sufficiently well informed so that they will not make mistakes. This is a strong assumption, which will permit some strong conclusions. We will discuss the limitations of this assumption, first in the context of the main results about equilibria, and then in Section 4.5, when we relax some of the assumptions about what the agents know, and discuss what are called games of incomplete information. However, some idea of what is involved in assuming that an equilibrium will be reached can be gotten from comparing equilibrium strategies with dominant strategies. This will give us some idea of the potential limitations of the (descriptive) assumption that equilibrium will be achieved, as well as of the limitations of equilibrium analysis as a means of giving advice to the players.

To say a particular strategy for some player is an *equilibrium strategy* (i.e., that it is an element of some equilibrium) is a much weaker statement than to say it is a dominant strategy. A dominant strategy is a best response to whatever other players may do, but an equilibrium strategy need only be a best response to the other strategies that comprise one particular equilibrium. So there are two potential limitations to the applicability of a theory of equilibrium. The first is that when more than one equilibrium exists, the players must somehow be able to coordinate their strategy choices, since a set of equilibrium strategies that do not all correspond to the same equilibrium may not be an equilibrium. Second, a great deal of information about the preferences of other players may be needed to compute an equilibrium. Again, note the contrast with dominant strategies. If some player i has a dominant strategy $Q(i)$, then no information is needed about other players' preferences to know how to give player i good advice: The best advice is to play $Q(i)$. However if i has no dominant strategy, then he or she needs to determine how to respond to the actions of other players. We will see that the assumption that players are well informed enough to arrive at an equilibrium may make serious demands on how well informed they need to be.

4.4.2   *Equilibrium behavior*

In this section, we consider the game induced by the M-optimal stable mechanism, and concentrate on those equilibria in which each man chooses his dominant strategy, and states his true preferences. According to the definition we have given of equilibrium, there may also be other equilibria. (Consider for example the preferences Q in which each agent states that all mates are unacceptable. Then all agents remain single, and this is an equilibrium, since no individual agent can change the outcome by changing his or her stated preferences while the other agents do not change theirs.) But the equilibria in which agents who have dominant strategies use them is a reasonable class of equilibria to examine because it is always one of an agent's best responses to choose a dominant strategy when one is available.

Having observed that truth telling is not an equilibrium, we need to ask whether equilibria always exist. Theorem 4.17 will show that they do. In Theorem 4.15 we first consider the case in which all preferences must be strict, and show that in this case there is an equilibrium set of strategies corresponding to each stable matching. (This theorem also will make very clear what we have been referring to when we say the players may need to be able to coordinate their strategy choices in order that an equilibrium should result, since when many stable matchings exist, the theorem describes many equilibria, but an equilibrium will result only when all players choose the equilibrium strategy corresponding to one stable matching.)

Theorem 4.15 (Gale and Sotomayor). *When all preferences are strict, let μ be any stable matching for (M, W, P). Suppose each woman w in $\mu(M)$ chooses the strategy of listing only $\mu(w)$ on her stated preference list of acceptable men (and each man states his true preferences). This is an equilibrium in the game induced by the M-optimal stable matching mechanism (and μ is the matching that results).*

Proof: It is clear that μ is stable under these stated preferences that we denote by P'. Further, μ is the only stable matching for (M, W, P'), for any other matching would leave some w in $\mu(M)$ unmatched, which by Theorem 2.22 is not possible at a stable matching for (M, W, P'). Hence μ is the M-optimal matching for (M, W, P').

By Theorem 4.7 we know truth telling is the best response of every man to any strategy choices of the other players. To see that P' is an equilibrium, suppose some w now changes her preference list, leading to a new set of stated preferences $P'' = (P'_{-w}, P'(w))$ with a new M-optimal matching μ' that gives her a mate $m' = \mu'(w)$ whom she prefers to $\mu(w)$ under her true preferences. Then m' must have been matched by μ to some w' who he prefers to w, for if not (m', w) would have blocked μ in (M, W, P). But then w' is self-matched under μ' since m' was the only acceptable man on her P'-list. So (m', w') blocks μ' under P'', a contradiction.

When preferences need not be strict, and an M-optimal stable matching mechanism with tie breaking is used, the strategies stated in Theorem 4.15 may not be in equilibrium, although equilibria nevertheless exist.

The impossibility theorem tells us that no stable matching mechanism exists that always makes truth telling by all the players an equilibrium. Theorems 4.15 and 4.17 tell us that some equilibria nevertheless exist when an M-optimal stable mechanism is employed, and that these equilibria involve misrepresentation by the women. This raises the question posed at the beginning of this chapter: In the light of this insight into the incentives facing the players, do we need to reevaluate our presumption in Chapter 2 that stable matchings will occur? That is, given that some women will typically have an incentive to misrepresent their preferences when an M-optimal stable mechanism is employed, do we have any reason to expect that the resulting matching will be stable with respect to the true preferences? If we expect equilibria to occur, the following theorem answers this question affirmatively.

Theorem 4.16 (Roth). *Suppose each man chooses his dominant strategy and states his true preferences, and the women choose any set of strategies (preference lists) $P'(w)$ that form an equilibrium for the matching game induced by the M-optimal stable mechanism. Then the corresponding M-optimal stable matching for (M, W, P') is one of the stable matchings of (M, W, P).*

Theorem 4.16 states that *any* equilibrium (in which the men state their true preferences) produces a matching that is stable with respect to the *true* preferences. When preferences are strict, this is a sort of converse to Theorem 4.15, which said that in that case any matching μ that is stable under the true preferences can be obtained by an equilibrium set of strategies.

Proof: Suppose μ'_M is the *M*-optimal matching for (M, W, P') but (m, w) blocks μ'_M under w's true preference. We will show that P' is not an equilibrium. Suppose that, instead of stating $P'(w)$, w lists only m on her preference list. Then she will get him, for let μ''_M be the *M*-optimal matching for the new preferences P'', which differ from P' only in w's preference list. If w does not get m, then she is single under the matching μ''_M, so by the stability of μ''_M, m prefers $\mu''_M(m)$ to w and by assumption m prefers w to $\mu'_M(m)$, so m prefers $\mu''_M(m)$ to $\mu'_M(m)$. But clearly the matching μ''_M would be stable for $(M, W-\{w\}; P')$, where P' is restricted to $W-\{w\}$. Thus m is worse off under the *M*-optimal matching for (M, W, P') than he is under μ''_M for $(M, W-\{w\}, P')$, which contradicts Theorem 2.25.

Another way to prove Theorem 4.16 is to consider the deferred acceptance procedure with men proposing. The key observation is the following: Let P' be any stated preferences (with the men stating the truth) such that μ'_M is blocked under the true preferences P by some pair (m, w). Then m must have proposed to w at some step of the algorithm (since $P'(m) = P(m)$), and w rejected him. So if w had submitted a preference $P''(w)$ listing m as her first choice, but otherwise agreeing with $P'(w)$, she would have been matched to m. So the instability of μ'_M implies that P' was not an equilibrium.

Theorem 4.15 showed that by misstating their preferences appropriately the women can obtain any stable matching by equilibrium strategies (when preferences are strict). Theorem 4.16, on the other hand, shows that the women cannot get too greedy, since if any set of strategies gives some woman w a mate whom she likes better than $\mu_W(w)$, this will not be an equilibrium.

By concentrating on equilibria, we have been considering sets of strategies with the property that no individual can profitably change his or her strategy. We can also use the strategic form of a game to look at the behavior of coalitions. We will say that a set of strategies forms a *strong equilibrium for the women* if it is an equilibrium and if no coalition of women can achieve a better payoff for all of its members by having its members change their strategies.

When preferences are strict, Theorem 4.15 says that the women can achieve any stable matching μ by some equilibrium set of strategies. However, unless $\mu = \mu_W$ the particular equilibrium strategies described in Theorem 4.15 will not be a strong equilibrium for the women. To see this, note that if $\mu \neq \mu_W$ then $\mu(w) \neq \mu_W(w)$ for at least two women, since if $\mu(w_1) \neq \mu_W(w_1) = m_1$, then by Theorem 2.22, m_1 is not single at μ and hence $\mu(m_1) = w_2$ and $\mu(w_2) = m_1 \neq \mu_W(w_2)$. To show μ is not a matching corresponding to a strong equilibrium for the women, let $\bar{W}$ be all w such that $\mu(w) \neq \mu_W(w)$. Let each w in $\bar{W}$ pretend $\mu_W(w)$ is her only acceptable man. Then μ_W is stable for these new preferences (in which all other agents state preferences as in Theorem 4.15) and since all w in $\bar{W}$ are matched by μ_W, they are matched by the M-optimal matching for the new preferences.

Once again, having defined a particular solution concept, we would like to know whether strong equilibria for the women always exist. The following theorem shows that they do. (For simplicity we state and prove the theorem as though preferences are strict, so optimal stable matchings exist, but if not, these can be understood as the matchings obtained after tie-breaking, as in our previous treatment of preferences that need not be strict.)

Theorem 4.17 (Gale and Sotomayor). *Let P' be a set of preferences in which each man states his true preferences, and each woman states a preference list that ranks the men in the same order as her true preferences, but ranks as unacceptable all men who are ranked below $\mu_W(w)$. These preferences P' are a strong equilibrium for the women in the game induced by an M-optimal stable matching mechanism (and μ_W is the matching that results).*

Proof: We claim μ_W is the only stable matching for (M, W, P'), for clearly μ_W is stable and any other stable matching μ' must have $\mu'(w) \neq \mu_W(w)$ for some w; hence w is not single under μ'. Then $\mu'(w)$ is preferred by w to $\mu_W(w)$ by construction of P'. Since μ_W is the W-optimal stable matching, this means μ' is unstable under the true preferences, hence it is blocked by some pair (m, w). But by the construction of P', this implies (m, w) blocks μ' under the preferences P', contradicting the P'-stability of μ'.

Now since μ_W is the only stable matching for (M, W, P') it is the W-optimal stable matching for P'. If some subset $\bar{W}$ could get a better payoff by stating different preferences, then it would get a better payoff than from the W-optimal matching of (M, W, P'), but by Theorem 4.10 and the symmetry between men and women this is not possible.

Theorem 4.17, for example, makes clear what we meant when we said that, quite apart from the problems of coordination among players that may arise in achieving some equilibrium set of strategy choices, a great deal of information may be required for players to be able to compute an equilibrium strategy. To compute the strategies described in the theorem, each woman needs to know $\mu_W(w)$, which can be computed only with a good deal of information about the preferences of the other men and women. This is information that would be difficult to come by in some of the environments to which we might hope to apply other aspects of the theory developed here. Consequently, theorems of this kind need to be interpreted with caution, a matter to which we return in Section 4.5.

In view of our earlier comments about the difficulties associated with coordination at a particular equilibrium, it would have been nice to be able to assert that the matching μ_W is the only matching obtainable from a strong equilibrium for the women. However the following example shows that this is not the case.

Example 4.18: Multiple strong equilibria (Gale and Sotomayor)
The true preferences follow.

$$P(m_1): w_2, w_1, w_3$$
$$P(m_2): w_1, w_3, w_2, w_4$$
$$P(m_3): w_3, w_4$$
$$P(m_4): w_4, w_3, w_1, w_2$$

$$P(w_1): m_4, m_1, m_2$$
$$P(w_2): m_4, m_2, m_1$$
$$P(w_3): m_4, m_3, m_2$$
$$P(w_4): m_4$$

For these preferences the optimal stable matching for the women is given by

$$\mu_W = \begin{matrix} w_1 & w_2 & w_3 & w_4 \\ m_1 & m_2 & m_3 & m_4 \end{matrix}.$$

Now suppose all women state the preferences, P', given below:

$$P'(w_1) = m_4, m_2$$
$$P'(w_2) = m_4, m_1$$
$$P'(w_3) = m_4, m_2, m_3$$
$$P'(w_4) = m_4.$$

The M-optimal matching for these new preferences is given by

$$\mu_M' = \begin{matrix} w_1 & w_2 & w_3 & w_4 \\ m_2 & m_1 & m_3 & m_4 \end{matrix}.$$

Although $\mu_M' \neq \mu_W$, we assert that P' is a strong equilibrium for the women. In fact, no subset W' of W that contains w_4 can, by stating different preferences, improve the situation of all its members, since the mate of w_4 is the best possible; if W' contains w_3, it cannot improve the situation of w_3, for if so (m_4, w_4) would block the new matching. If W' contains w_2, it cannot get a better mate for w_2 by stating different preferences, for if so (m_4, w_4) or (m_2, w_3) would block the new matching. If W' contains w_1, it is not possible to improve the payoff of w_1, for if so (m_4, w_4) or (m_1, w_2) would block the new matching.

4.4.3   Good and bad strategies

Assuming for the moment that the women could get the information needed to know how to misrepresent their preferences as in Theorem 4.17, there is a sense in which we could consider the strategies $P'(w)$ described in that theorem as the best method of play for the women. These strategies give rise to a strong equilibrium for the women, with the further property that among all equilibria (at which men choose their dominant strategy of stating their true preferences), it gives the women their most preferred matching.

But it is clear that advising a woman to choose this strategy will be singularly unhelpful if she does not know $\mu_W(w)$, and so perhaps we should think of it as a "good strategy" only in environments in which all the players know one another's preferences. This leads us to consider what advice we can give in environments in which information about other players' preferences may not be readily available to the players.

We have already observed that problems of coordination and information do not arise in the same way for players who have a dominant strategy. In particular, Theorem 4.7 implies that when an M-optimal stable matching procedure is used, a man may confidently state his true preferences, without regard to what the preferences of the other men and women may be. So this is a good strategy for the men, and other strategies are, in comparison, bad. Although we have seen that stating the true preferences is not a good strategy in the same way for the women, we turn now to considering what classes of strategies might be bad, in the sense of being dominated by other available strategies.

Following our definition of a *dominant strategy,* we can say that a strategy $Q(i)$ for some player i *dominates* another strategy $Q'(i)$ for the same player if the payoff to i when he or she plays $Q(i)$ is at least as high as when he or she plays $Q'(i)$, no matter what strategies the other players play. We now want to sharpen this notion a little, as follows.

Definition 4.19. *A strategy $Q(i)$ for player i strictly dominates another strategy $Q'(i)$ if $Q(i)$ dominates $Q'(i)$ and if, for at least one set of strategies Q_{-i} for the other players, i prefers the outcome when he or she plays $Q(i)$ to the outcome when he or she plays $Q'(i)$.*

To simplify the discussion, we will state and prove the results in the remainder of this section as if all women have strict preferences. However each of the results applies as well when preferences need not be strict, but wherever we speak of a woman's first choice, it should be understood that we mean all of the men who are tied for first in her preferences.

If there is only one woman, w, then revealing her true preferences strictly dominates her other strategies, for if she submits her true preferences, she will get the highest man on her list who has listed her. If she were to instead state $m' >_w m$ when in fact $m >_w m'$, she would be worse off in the case where m and m' are the only men who are willing to marry her. From now on it will be assumed that there are at least two women in W. We will show that the strategy of listing only one man, as in Theorem 4.15, is dominated unless that man happens to be the women's true first choice. In fact, we can essentially characterize the dominated strategies of the women by the following two theorems.

Theorem 4.20 (Roth). *Any strategy $P'(w)$ in which w does not list her true first choice at the head of her list is strictly dominated, in the game induced by the M-optimal stable mechanism.*

Proof: Let $P'(w)$ be such a strategy. Let m_1 be w's true first choice. We will show that $P'(w)$ is strictly dominated by $P''(w)$ that lists m_1 in first place and leaves the rest of $P'(w)$ unchanged. Let μ'_M and μ''_M be the corresponding M-optimal matchings (the strategies of all other players are fixed). First suppose $\mu''_M(w) \neq m_1$. Then $\mu''_M(m_1) >_{m_1} w$ for if not (m_1, w) would block μ''_M. Hence μ''_M is stable under preference $P'(w)$. So,

$\mu'_M(m_1) \geq_{m_1} \mu''_M(m_1)$, from the M-optimality of μ'_M. These two conclusions imply $\mu'_M(m_1) >_{m_1} w$, and since all the other elements different from m_1 are ranked in the same ordering in both lists $P'(w)$ and $P''(w)$, it follows that μ'_M is stable under $P''(w)$ and hence $\mu''_M(w) = \mu'_M(w)$; so w is no worse off using $P''(w)$ than using $P'(w)$. (If $\mu'_M(w) = m_1$, this is immediate.) To show that $P''(w)$ strictly dominates $P'(w)$, let m' be the first element in $P'(w)$. Consider the following representation of preferences: $P(m') = w$, $P(m_1) = w$, and no other man lists w. Then we can see that $m_1 = \mu''_M(w) >_w \mu'_M(w) = m'$, which concludes the proof.

Theorem 4.21 states that Theorem 4.20 describes essentially all the dominated strategies.

Theorem 4.21 (Gale and Sotomayor). *Let $P'(w)$ be any strategy for w in which (a) w's true first choice is listed first, and (b) the acceptable men in $P'(w)$ are also acceptable men in w's true preference list $P(w)$. Then $P'(w)$ is not a dominated strategy when the M-optimal stable mechanism is used.*

Proof: We will show that for any other strategy $P''(w)$ there exist strategies P_{-w} for the other players such that $\mu'_M(w) >_w \mu''_M(w)$, where μ'_M and μ''_M are the M-optimal matchings for $(M, W; (P'(w), P_{-w}))$ and $(M, W; (P''(w), P_{-w}))$, respectively. Let m_1 be w's true first choice. There are three cases. We first suppose $P''(w)$ also satisfies assumption (a) above.

Case 1: m is an acceptable man in $P'(w)$ but not in $P''(w)$. Then for P_{-w} we suppose that m lists w as first choice and no other man lists w. Then $\mu'_M(w) = m$ while w is self-matched under μ''_M, so w prefers μ' (here we use the fact that m is acceptable under the true preferences $P(w)$).

Case 2: m is an acceptable man in $P''(w)$ but not in $P'(w)$. Then for P_{-w} suppose $P_{-w}(m) = w, w'$, for some w'; $P_{-w}(m_1) = w', w$; $P_{-w}(w') = m, m_1$ and no other man lists w or w'. Then $\mu'_M(w) = m_1 >_w \mu''_M(w) = m$.

Case 3:   Let $P'(w)$ and $P''(w)$ contain the same set of acceptable men but w prefers m to m' in $P'(w)$ and m' to m in $P''(w)$. Then suppose $P_{-w}(m_1) = w', w$; $P_{-w}(m) = w, w'$; $P_{-w}(m') = w, w'$; and $P_{-w}(w') = m', m_1, m$; and no other men list either w or w'. It is an instructive exercise to verify that $\mu'_M(w) = m_1$, while $\mu''_M(w) = m'$.

We have seen that if $P'(w)$ satisfies (a) and (b) and $P''(w)$ satisfies (a), then for some P_{-w}, $\mu'_M(w) >_w \mu''_M(w)$. If $P''(w)$ does not satisfy (a), then by Theorem 4.20 there is some $P'''(w)$ that dominates $P''(w)$ and $P'''(w)$ satisfies (a) so we construct P_{-w} so that $\mu'_M(w) >_w \mu'''_M(w)$ but $\mu''_M(w) \geq_w \mu'''_M(w)$ by dominance. The proof is now complete.

Remark 1: Condition (b) of Theorem 4.21 is needed to avoid cases in which $P'(w) = m_1, m$, where m is not an acceptable man in $P(w)$. This is strictly dominated by $P''(w) = m_1$.

Remark 2: Note that Example 4.18 uses only undominated strategies for the women. Thus we do not have a unique strong equilibrium even when players are restricted to use only undominated strategies.

4.5 Incomplete information about others' preferences

We have so far avoided making explicit assumptions about precisely what the agents know. Nevertheless, as we have seen in our exploration of equilibrium in Section 4.4.2, strong implicit assumptions about what the agents know reveal themselves in the equilibrium strategies. The precise reasons for this would take us too far afield for our present purposes, but in recognition of the implicit assumptions about information contained in the equilibrium analysis of games as we have so far described them, such games are referred to in the literature as games of *complete information*.

The customary assumption about games of complete information is that all the information in the model used to analyze the game is *common knowledge* among the players. (A piece of information is common knowledge when all of the arbitrary-length "sentences" of the form "Player i knows that j knows that i knows that j knows...the information" are true. So an event becomes common knowledge among a group of people if it occurs in public, so not only do they all see it, but they see each other seeing it, and see that everyone sees that they see it, and so forth.) Among the consequences of assuming that the strategic form model ($N = M \cup W$, $\{D_i\}, h, P$) we have been considering is common knowledge among the players is that it implies that the players know one another's preferences, since these are part of the model. This has perhaps been clearest in Theorem 4.17, in which each woman's equilibrium strategy requires her to be able to compute the matching μ_W.

Since the common knowledge assumption is such a strong one, we need to be able to relax it in order to model many important situations. Games in which some features of the game are not common knowledge among the players are called *games of incomplete information*. We turn next to consider them.

Consider a model in which each agent does not know the preferences of the others, but knows only the probability distribution from which these preferences are drawn. Since we will be dealing with probabilities, we will need to consider not merely the preference orderings of the players, but also their expected utility functions. (Preferences will still be assumed to be strict.) We will also consider games with quite general rules and strategy sets, and not merely revelation games in which agents simply state their preferences. For example, the rules considered at the very beginning of this chapter, in which men make proposals and women accept or reject them, define a game in which the decisions facing the players do not involve stating any preferences.

We begin with a few words about expected utility functions. Up to this point we have only been concerned with the simple preferences of the agents, of the kind that can be represented by preference orderings, to indicate that man m_1, say, prefers w_1 to w_2 and w_2 to w_3. But a more detailed description of preferences is needed if we wish to know, for example, if man m_1 would prefer to be matched to w_2, or to face a lottery in which he would be matched to w_1 with probability p, and to w_3 with probability $1-p$. It is this kind of additional information that an expected utility function is meant to represent.

An *expected utility function* for man m_i is a real-valued function u_{m_i} defined on the set $W \cup m_i$ of possible matches that m_i could receive, with the property that for all a, b in $W \cup m_i$,

$$u_{m_i}(a) \geq u_{m_i}(b) \quad \text{if and only if} \quad a >_{m_i} b;$$

and for any probability p and lottery $[pa; (1-p)b]$ [which yields alternative a with probability p and b with probability $(1-p)$], the utility of the lottery is given by its expected utility, that is,

$$u_{m_i}[pa; (1-p)b] = pu_{m_i}(a) + (1-p)u_{m_i}(b).$$

A large literature has been devoted to determining what kinds of preferences can be faithfully represented in this way. However in this section we will assume without further ado that an agent with expected utility function u prefers a lottery $[pa; (1-p)b]$ to another alternative c (which may or may not be a lottery itself) if and only if $u([pa; (1-p)b]) > u(c)$. Note that the set U_i of all expected utility functions that agent m_i, say, could have is much larger than the (finite) set of all possible preference orderings he could have over possible matches, since an expected utility function reflects not only the simple *order* of his preferences over possible mates, but also a measure of their intensity, as measured by the risks he is willing to take to achieve different matches.

A general matching game with incomplete information about others' preferences will be given by a collection

$$\Gamma = (N = M \cup W, \{D_i\}_{i \in N}, g, U = \times_{i \in N} U_i, F).$$

The set N of players consists of the men and women to be matched. The sets D_i describe the decisions facing each player in the course of any play of the game (i.e., an element d_i of D_i specifies the action of player i at each point in the game at which he or she has decisions to make). The function g describes how the actions taken by all the agents correspond to matchings and lotteries over matchings, that is, $g: \times_{i \in N} D_i \to L[M]$, where M is the set of all matchings between the sets M and W, and $L[M]$ is the set of all (finite) probability distributions (lotteries) over M. The set U_i is the set of all expected utility functions defined over the possible mates for player i and the possibility of remaining single, and F is a probability distribution over n-tuples of utility functions $u = \{u_i\}_{i \in N}$, for u_i in U_i. The interpretation is that a player's "type" is given by his or her utility function, and at the time players must choose their strategies, each player knows his or her own type, and the probability distribution F over vectors u is common knowledge. The special case of a game of complete information occurs when the distribution F gives a probability of one to some vector u of utilities. Whereas the set U of possible utilities for the players is large, we will typically be concerned with games in which only a subset of U has positive probability. For simplicity, we will henceforth confine our attention to cases in which the set of utility functions that occur with positive probability is countable. In any event, since each player i knows his or her own utility function u_i, he or she can compute a conditional probability $p_i(u_{-i} | u_i)$ for each vector of other players' utilities u_{-i} in $U_{-i} \equiv \times_{j \neq i} U_i$, by applying Bayes' rule to F.

This isn't the most general kind of incomplete information model we might consider (cf. Harsanyi 1967, 1968a, b). In particular, players know their own utilities for being matched with one another even though they do not know what "type" the other is. The final matching, and hence each player's utility payoff, depends only on the actions of the players, not on their types. (That is, each player's utility payoff depends on his or her own type, and on the actions of all the players, but not on the types of the other players.) To put it another way, players' types don't affect their desirability, only their desires. This seems like a natural assumption for elite professional markets for entry level positions. For example in the hospital intern market, after the usual interviewing has been completed, top students are able to rank prestigious programs, and vice versa. But agents don't know how their top choices rank *them*. (Note the difference between this kind of model and one in which the interviewing process itself is modeled, in which agents would in effect be uncertain about their own preferences.)

A *strategy* for player i is a function σ_i from his or her type (which in this case is his utility function) to his or her decisions, that is, $\sigma_i: U_i \to D_i$. If $\sigma = \{\sigma_i\}_{i \in N}$ denotes the strategy chosen by each player, then for each vector u of players' utility functions, $\sigma(u) = \{d_i \in D_i\}_{i \in N}$ describes the decisions made by the players, which result in the matching (or lottery over matchings) $g(\sigma(u))$. Consequently a set of strategy choices σ results in a lottery over matchings, whose probabilities are determined by the probability distribution F over vectors u, and by the function g. The expected utility to player i who is of type u_i is given by

$$u_i(\sigma) = \sum_{u_{-i} \in U_{-i}} p_i(u_{-i}|u_i) u_i[g(\sigma(u_{-i}, u_i))].$$

An *equilibrium* set of strategies is a σ^* such that for all players i in N and all utility functions u_i in U_i, $u_i(\sigma^*) \geq u_i(\sigma^*_{-i}, \sigma_i)$ for all other strategies σ_i for player i. That is, when player i's utility is u_i the strategy σ_i^* determines player i's decision $d_i^* = \sigma_i^*(u_i)$, and the equilibrium condition requires that for all players i and all types u_i that occur with positive probability, player i cannot profitably substitute another decision $d_i = \sigma_i(u_i)$.

4.5.1   Revelation games

We turn now to consider the class of incomplete information games called *revelation games,* in which players are called upon only to state (*reveal*) their types, which in this case are their utilities. (We will later also consider the special case of revelation games in which players can only state preference orderings, and not numerical utilities. This corresponds both to what we have considered in the special case of complete information, and also to the situation in the actual two-sided matching markets discussed in Sections 1.1, 5.4, and 5.5 that employ revelation procedures.) Recall that a general matching game with incomplete information about others' preferences is given by $\Gamma = (N = M \cup W, \{D_i\}_{i \in N}, g, U = \times_{i \in N} U_i, F)$. We may call $[\{D_i\}_{i \in N}, g]$ the *mechanism,* and $[U, F]$ the *state of information* of the game. Then a game Γ is specified by a set of players, a mechanism, and a state of information. A *revelation game* Γ_R is a game in which the mechanism is of the form $[\{D_i = U_i\}_{i \in N}, h]$, where h is a function that takes stated preferences into matchings or lotteries over matchings, that is, $h: \times_{i \in N} D_i \to L[M]$. We will sometimes call the function h itself a *revelation mechanism,* it being understood that the decision facing each agent is simply what utility to state. We will pay particular attention to the strategy of truth telling in a revelation game, that is, the strategy $\sigma_i^T(u_i) = u_i$ in which each agent always states his or her true type.

For any general game Γ of incomplete information, and any equilibrium σ^* of Γ, we can define the *revelation game corresponding to* σ^* to be the game $\Gamma_R(\sigma^*)$ with the same set of players and state of information as Γ and with the revelation mechanism h given by $h(u) = g(\sigma^*(u))$. That is, the revelation mechanism h takes any set of stated utilities and produces the same matching (or lottery over matchings) that would have been produced by the equilibrium σ^* in the game Γ if the true utilities of the players had been u. (That is, regardless of the actual types of the players in the corresponding revelation game, if they collectively state the utilities u, the resulting matching is the one that would have resulted in Γ under strategies σ^* if the vector u corresponded to the true player types.) The following observation, which is widely used in proofs about games of incomplete information, has become known as the "revelation principle."

The revelation principle. *For any equilibrium σ^* of a general incomplete information matching game Γ, let $\Gamma_R(\sigma^*)$ be the corresponding revelation game. Then*

1. *Truth telling is an equilibrium. That is, the strategies $\sigma^T = \{\sigma_i^T\}_{i \in N}$ are an equilibrium in $\Gamma_R(\sigma^*)$.*
2. *When all players tell the truth in $\Gamma_R(\sigma^*)$, the resulting matching (or lottery over matchings) is the same as when the players play the strategies σ^* in Γ.*

The second observation follows immediately from the definition of the revelation mechanism h, and does not depend on whether or not the strategies σ^* are an equilibrium in the original game Γ. The first observation, that truth telling is an equilibrium in the revelation game, follows from the fact that σ^* is an equilibrium in Γ: If player i with utility u_i could profit from stating another utility function v_i when all other agents state the truth in the revelation game, then he or she could get the same outcome in game Γ, and hence also profit, from playing $\sigma(v_i) \equiv d_i$ instead of $\sigma^*(u_i) \equiv d_i^*$ when all other players use the strategies σ^*. But then σ^* would not be an equilibrium, contrary to assumption.

We will call a revelation mechanism h *stable* if for any stated utilities u its output $h(u)$ is a stable matching or lottery over stable matchings, that is, if $h(u) \in L[S(u)]$. Note that the set of stable matchings is sensitive only to the ordinal preferences; that is, if $u = (u_1, \ldots, u_{n+p})$ is a vector of expected utility functions, one for each agent, then there is a unique vector $P = P(u)$ of ordinal preferences corresponding to these utilities, and the set of stable matchings is the same for any two utility vectors that have the same corresponding preferences, that is, $S(u) = S(v) = S(P)$ whenever $P(u) = P(v) = P$.

Of course, while the set of stable matchings responds only to the ordinal information contained in the expected utility functions, a mechanism for selecting a stable matching can depend on the expected utilities in a more detailed way. In the complete information case we considered mechanisms that respond only to stated preferences. (As mentioned earlier, this corresponds to what is generally observed in markets that employ centralized matching mechanisms, such as the market for American medical interns.) The M-optimal and W-optimal stable mechanisms, for example, only depend on the preferences, not on the utilities. Formally, with respect to a set $M \cup W$ of agents we can define the following special class of revelation mechanisms.

Definition 4.22. *An ordinal (and nonrandom) stable matching mechanism is a function defined on all utility vectors u such that $h(u)$ is in $S(P)$ where P are the preferences corresponding to u, and such that if v is another utility vector corresponding to P, then $h(u) = h(v)$.*

Note that when an ordinal mechanism is used, it is equivalent to think of agents as stating either their utility functions or the corresponding preferences. In contrast to ordinal stable mechanisms, the general stable revelation mechanisms we have defined could be called *random cardinal* stable mechanisms, in recognition of the fact that they may yield different (and random) outcomes for stated utilities u and v for which $P(u) = P(v)$. For example, we might consider the random mechanism that yields μ_M or μ_W each with probability one-half, or a cardinal mechanism that chooses the optimal stable matching of whichever side of the market has the agent who states the highest utility number. (This latter mechanism has very unattractive incentive properties.)

4.5.2   *Equilibrium and stability*

The first result is an impossibility theorem that provides a strong negation to the conclusions of Theorem 4.16 about equilibria in the complete information case when the M-optimal stable mechanism is employed. It says that in the incomplete information case, no equilibrium of any mechanism can have the stability properties that every equilibrium of the M-optimal stable mechanism has in the complete information case. The strategy of the proof will be to observe that by the revelation principle, if any such mechanism existed then there would be a stable revelation mechanism with truth telling as an equilibrium, and then to show that no such revelation mechanism exists. The proof of Theorem 4.23 thus also shows that the impossibility result of Theorem 4.4 generalizes to the case of cardinal mechanisms and incomplete information.

Theorem 4.23 (Roth).   *If there are at least two agents on each side of the market, then for any general matching mechanism $[\{D_i\}_{i \in N}, g]$ there exist states of information $[U, F]$ for which **every** equilibrium σ of the resulting game Γ has the property that $g(\sigma(u)) \notin L[S(u)]$ for some $u \in U$. (And the set of such u with $g(\sigma(u)) \notin L[S(u)]$ has positive probability* *under F.) That is, there exists no mechanism with the property that at least one of its equilibria is always stable with respect to the true preferences at every realization of a game.*

To see that at least two agents are required on each side of the market, note that since preferences are strict, a game with only one agent on one side of the market must have a unique stable matching, at which the singleton agent gets his or her highest ranked mutually acceptable choice. In such a game it is not hard to see that it is a dominant strategy for all agents to state their true preferences when a stable mechanism is used.

The proof will look at the smallest remaining case, of two agents on each side of the market, and show an example, that is, a state of information, that causes every stable revelation mechanism (and hence every mechanism) to fail. The conclusion for larger sets of agents follows from the fact that the four agents who play a role in the proof can be embedded in any larger set of agents without affecting the conclusion, so long as their preferences are not changed (and so that, in particular, in a larger market these four agents do not consider any additional agents to be acceptable matches).

The example
In order not to obscure the basic simplicity of the proof, it will be helpful to state the ordinal preferences of the agents first, and consider their utility functions later. The agents are $N = M \cup W$, where $M = \{m_1, m_2\}$ and $W = \{w_1, w_2\}$. The most likely distribution of player types corresponds to the following preferences.

$$P(m_1) = w_1, w_2 \qquad P(w_1) = m_2, m_1$$
$$P(m_2) = w_2, w_1 \qquad P(w_2) = m_1, m_2$$

Agents m_2 and w_2 have no other types that occur with positive probability, but m_1 and w_1 may each have two types, with their other possible preferences being

$$P'(m_1) = w_1 \qquad P'(w_1) = m_2.$$

The probability that m_1 has preferences $P'(m_1)$ rather than $P(m_1)$ is q, which is also the probability that w_1 has preferences $P'(w_1)$ rather than $P(w_1)$. Let

$$P = (P(m_1), P(m_2); P(w_1), P(w_2)),$$
$$P' = (P'(m_1), P(m_2); P(w_1), P(w_2)),$$
$$P'' = (P(m_1), P(m_2); P'(w_1), P(w_2)), \quad \text{and}$$
$$P''' = (P'(m_1), P(m_2); P'(w_1), P(w_2))$$

be the various preference profiles that can arise with probabilities $(1-q)^2$, $q(1-q)$, $q(1-q)$, and q^2, respectively.

We will suppose that for each type of each agent, the utility of being matched to his or her first choice is 2, to his or her second choice is 1, and to his or her third choice 0. Thus, for example, when m_1 has preferences $P(m_1)$ his utility is $u_{m_1}(w_1) = 2$, $u_{m_1}(w_2) = 1$, and $u_{m_1}(m_1) = 0$, and when he is the type with preferences $P'(m_1)$ his utility is given by $u'_{m_1}(w_1) = 2$, $u'_{m_1}(m_1) = 1$, $u'_{m_1}(w_2) = 0$.

There are three distinct matchings that may be stable for some realization of the possible types. Denote these by

$$\mu = \begin{pmatrix} m_1 & m_2 \\ w_1 & w_2 \end{pmatrix}, \qquad \nu = \begin{pmatrix} m_1 & m_2 \\ w_2 & w_1 \end{pmatrix}, \qquad \tau = \begin{pmatrix} m_1 & m_2 & (w_1) \\ (m_1) & w_2 & w_1 \end{pmatrix}.$$

The sets of stable matchings corresponding to each of the possible combinations of types are

$$S(P) = \{\mu, \nu\}; \qquad S(P') = \{\mu\};$$
$$S(P'') = \{\nu\}; \qquad S(P''') = \{\tau\}.$$

Proof of Theorem 4.23:   Notice that this example is symmetric between the two sides of the market. Since $S(P)$ contains only two matchings, we may suppose without loss of generality that an arbitrary stable revelation mechanism h chooses the stable matching μ with probability at least one-half when the utilities u corresponding to the preferences P are stated. (Since there are only two stable matchings in this case, a stable mechanism must choose one of them with a probability of at least one-half. If h instead chose ν with probability greater than one-half, the argument that follows would proceed with m_2 replacing w_2.) We will show that h gives w_2 an incentive to misstate her utility when the other agents adopt the strategy of truth telling, and therefore that truth telling is not an equilibrium for any stable revelation mechanism when the state of information is as given in the example.

Let σ be a strategy 4-tuple, and denote by $\sigma \mid P$ the vector of stated utilities that results when the true types of the players correspond to the preferences P, and similarly denote by $\sigma \mid P'$, $\sigma \mid P''$, and $\sigma \mid P'''$ the stated utilities that correspond to the other configurations of player types. Then the expected utility of w_2, whose utility function is u_{w_2} as specified earlier is

$$u_{w_2}(\sigma) = (1-q)^2 u_{w_2}(h(\sigma \mid P))$$
$$+ q(1-q)[u_{w_2}(h(\sigma \mid P')) + u_{w_2}(h(\sigma \mid P''))] + q^2 u_{w_2}(h(\sigma \mid P''')).$$

So, if q is sufficiently small, any potential losses in states of the world other than P are offset in expected utility by any gains in the most probable state of the world, P. Thus for $q = 0.01$, say, w_2 prefers to get ν for certain instead of μ with a probability of at least one-half in state P, at the cost of getting any other match in any of the other states. But w_2 can achieve this when all other agents adopt the truth-telling strategy, by stating any utility function corresponding to the preferences $P'(w_2) = m_1$, which are not her true preferences. The reason is that when all other agents state utilities corresponding to P, but w_2 states a utility corresponding to $P'(w_2)$, then the unique stable matching is ν, and so any stable mechanism must choose ν in this case. So truth telling is not an equilibrium, since when other agents adopt truth-telling strategies, w_2 will prefer to state preferences corresponding to $P'(w_2)$. (Since w_2 has only one type with positive probability, this fully describes her strategy, as far as it affects any agent's expected utility.)

By the revelation principle, the proof is now complete, since if any mechanism existed with an equilibrium that always produced a stable outcome, the corresponding revelation mechanism would be a stable mechanism in which truth telling was an equilibrium.

The next theorem shows that the conclusion of Theorem 4.10 also does not generalize to the case of incomplete information. It is possible for coalitions of men, by misstating their preferences, to obtain a preferable matching (even) from the M-optimal stable mechanism. This is so even though, as we will see in Theorem 4.26, it remains a dominant strategy for each man to state his true preferences.

Theorem 4.24 (Roth). *In games of incomplete information about preferences, the M-optimal stable mechanism may be group manipulable by the men.*

Proof: Consider the following example, which is adapted from Example 2.31 by the addition of agent w_4.

The agents are $N = M \cup W$, where $M = \{m_1, m_2, m_3\}$ and $w = \{w_1, w_2, w_3, w_4\}$. The most likely distribution of player types corresponds to the following preferences:

$$P(m_1) = w_4, w_2, w_1, w_3 \qquad P(w_1) = m_1, m_2, m_3$$
$$P(m_2) = w_4, w_1, w_2, w_3 \qquad P(w_2) = m_3, m_1, m_2$$
$$P(m_3) = w_1, w_2, w_3 \qquad P(w_3) = m_1, m_2, m_3$$
$$P(w_4) = w_4.$$

Note that under these preferences, woman w_4 is unwilling to be matched with any man. Except for w_4, each agent has only one type having positive probability, corresponding to the given preferences. Woman w_4 has two types having positive probability, with her other possible preferences being $P'(w_4) = m_1, m_2$.

The (small) probability that w_4 has preferences $P'(w_4)$ rather than $P(w_4)$ is q. Let $P = (P(m_1), P(m_2), P(m_3), P(w_1), \ldots, P(w_4))$ and $P' = (P(m_1), \ldots, P'(w_4))$ denote the two preference profiles that can arise, with probabilities $(1-q)$ and q, respectively.

Let $\mu_M(P)$ and $\mu_M(P')$ denote, respectively, the M-optimal stable matchings with respect to P and P'. Then

$$\mu_M(P) = \begin{pmatrix} m_1 & m_2 & m_3 & (w_4) \\ w_1 & w_3 & w_2 & w_4 \end{pmatrix} \quad \text{and} \quad \mu_M(P') = \begin{pmatrix} m_1 & m_2 & m_3 & (w_3) \\ w_4 & w_1 & w_2 & w_3 \end{pmatrix}.$$

For q sufficiently small (specifically, for $q < [u_{m_1}(w_2) - u_{m_1}(w_1)]/[u_{m_1}(w_4) - u_{m_1}(w_1)]$), the coalition of m_1 and m_2 can assure themselves a higher expected utility by stating the preferences $Q(m_1) = w_2, w_1, w_3$ and $Q(m_2) = w_4, w_3$ when other agents all state their true preferences. This is because

$$\mu_M(Q|P) = \begin{pmatrix} m_1 & m_2 & m_3 & (w_4) \\ w_2 & w_3 & w_1 & w_4 \end{pmatrix} \quad \text{and}$$

$$\mu_M(Q|P') = \begin{pmatrix} m_1 & m_2 & m_3 & (w_3) \\ w_2 & w_4 & w_1 & w_3 \end{pmatrix}.$$

Note that both men m_1 and m_3 profit from m_2's misrepresentation when the true preferences are P, and that m_2 gets the same spouse as if he had stated his true preferences in this case. But m_2 does better when the preferences are P' than he would have if m_1 had stated his true preferences; and although m_1 does worse in this case, he nevertheless receives a higher expected utility when he and m_2 both misstate their preferences according to Q, since the probability q is small. Note that when the time comes to state their preferences, neither m_1 nor m_2 knows whether the true preferences are P or P'.

As discussed in Section 4.3.1, the fact that even in the case of complete information it is possible for a coalition of men to misstate their preferences in a way that does not hurt any of them and helps some of them, means that the conclusion from Theorem 4.10 that coalitions of men cannot collectively manipulate the M-optimal mechanism to their advantage cannot be expected to be very robust. Once there is any possibility that the men can make any sort of side payments among themselves, this conclusion is no longer justified. Theorem 4.24 shows that uncertainty about the preferences of other agents allows some transfers in an expected utility sense, with men able to trade a gain in one realization for a gain in another. Note that this is so for any positive q, that is, even when q is arbitrarily small, in which case there is very little uncertainty about the preferences.

4.5.3 *Dominant and dominated strategies*

In contrast to the results for equilibria, the results concerning dominant strategies in the complete information case do generalize to the case of incomplete information. We begin with a general proposition about the relationship of dominant strategies for complete information revelation games and corresponding revelation games of incomplete information about others' preferences. (Here "corresponding" means having the same mechanism and set of players, but allows for different states of information, i.e., different probability distributions F over possible utilities.)

Proposition 4.25 (Roth).   *If h is a mechanism that makes stating his or her true utility u_i a dominant strategy for player i in every complete information revelation game Γ (i.e., for every specification of u), then the truth-telling strategy $\sigma_i^T(u_i) = u_i$ is a dominant strategy in any corresponding game Γ^* of incomplete information about others' preferences.*

Proof: Suppose not. Then there is another strategy σ_i that does better for at least some realization $u \in U$, such that $\sigma_i(u_i) \neq u_i$. But this contradicts the fact that in the complete information game with the utilities given by u, it is a dominant strategy for agent i to state u_i.

The following is an immediate consequence of Proposition 4.25 and Theorem 4.7.

Theorem 4.26 (Roth).   *In matching with incomplete information about others' preferences, the M-optimal stable mechanism makes it a dominant strategy for each man to state his true preferences; that is, $\sigma_i^T(u_i) = u_i$ is a dominant strategy for each man. (Similarly, the W-optimal stable mechanism makes it a dominant strategy for every woman to state her true preferences.)*

A similar, pointwise argument on realizations of the types of the players allows us to prove the following parallel to Theorem 4.20.

Theorem 4.27 (Roth).   *When an M-optimal stable mechanism is used in matching with incomplete information about others' preferences, any* strategy $\sigma_i(u_i)$ *for a woman* w_i *is dominated if her stated first choice is not her true first for each* u_i *in* U_i.

Proof: Consider a strategy σ_i for some woman w_i that for at least one u_i in U_i states a utility $\sigma_i(u_i)$ that ranks highest some alternative different from the highest ranked alternative according to u_i; that is, such that the maximum of $\sigma_i(u_i)$ over $M \cup \{w_i\}$ is achieved by some alternative other than the one that maximizes u_i, which we will denote s^*. Let $\sigma_i^*: U_i \to D_i$ be a strategy that differs from σ_i only for $u_i \in U_i$. Furthermore, suppose that the numbers $\sigma_i^*(u_i)(s) = \sigma_i(u_i)(s)$ for all $s \neq s^*$ in $M \cup \{w_i\}$, and that s^* maximizes $\sigma_i^*(u_i)$. Recall that the matching that results, and the corresponding utility that w_i derives, depends only on the stated utilities of the players (their actions) and not on their types. So by Theorem 4.20 woman w_i does at least as well by playing σ_i^* as she does by playing σ, for any strategy choices of the other agents, and does strictly better for at least one set of other agents' strategy choices.

Since the M-optimal stable mechanism is an ordinal mechanism, no difficulty in the proofs arises from the fact that Theorem 4.7 and 4.20 are stated in terms of a game in which agents state preferences, whereas Theorems 4.26 and 4.27 are stated in terms of a game in which agents state utilities.

Proposition 4.25 and the related argument in the proof of Theorem 4.27 illustrate a kind of "dominant strategy principle," connecting dominant strategy results for classes of complete information games to parallel results for incomplete information games, as exemplified by Theorems 4.26 and 4.27. As we have seen, no such parallels exist for the equilibrium results for the complete information game.

4.5.4   *Discussion of the incomplete and complete information models*

Since the questions in this section are motivated to large extent by empirical questions concerning the behavior of agents in real markets, it seems appropriate to conclude with some comments on modeling issues. The first of these concerns the definition of stability, which was carried over unchanged from the complete information model to the incomplete information model. In the context of the incomplete information model, the kind of stability studied here is ex post stability, in the sense that a stable matching would remain stable even if all the preferences were to become common knowledge.

The reason this is not an excessively strong requirement is that when a matching is proposed, each agent knows which alternatives he or she prefers. Although the agent does not yet know how his or her preferred alternatives evaluate him or her, the ease of ascertaining this in a number of the markets of interest is precisely what causes the instability observed in those markets. For example, in the hospital intern labor market, a medical student who has been offered an internship at his or her third choice hospital can easily contact the first and second choices to see if they prefer him or her to any of the students they are presently considering. In the late 1940's, prior to the introduction of a stable matching procedure in this market, just this kind of very late search caused verbal and other contracts to be broken in substantial enough numbers to interfere with the operation of the market (see Roth 1984a).

A larger modeling issue is the question of when complete and incomplete information models are most useful. It seems clear that in most of the markets to which these models can be applied, agents do not know with precision the preferences of all the other agents. However it is rarely apparent what, if any, priors about agents' preferences can be reasonably described as being shared by all agents. So both kinds of models impose costs. For certain kinds of questions about stability, its ex post nature makes the differences between the two kinds of models unimportant. For other questions, there seems to be no practical alternative to examining both kinds of models and the answers they give, in the light of how much they seem to make strained assumptions about the markets in question. In just this way, the information required to implement the equilibria identified by the complete information model led to the present exploration of the incomplete information case. However the equilibrium results obtained here are negative, and it remains an open question what general positive characterizations of equilibria can be obtained in the incomplete information case.

4.6   Guide to the literature

A number of the results presented in this section have been treated by several authors, who have often succeeded in extending earlier results and simplifying earlier proofs. Consequently many of the results must be attributed to more than one source.

The impossibility theorem (Theorem 4.4) was stated and proved in Roth (1982a). An alternate proof was subsequently developed independently by Theodore Bergstrom and Richard Manning (personal communication). The original proof in Roth (1982a) used three men and women, and showed that the result follows even if agents can only misrepresent their preference ordering but cannot misrepresent who is acceptable and unacceptable. Our proof of Theorem 4.6 required only a slight modification of the proof of Corollary 4.12 presented by Gale and Sotomayor (1985b).

Theorem 4.7 is from Roth (1982a). Lemmas 4.8 and 4.9 appeared there as part of the proof. Theorem 4.7 was also independently observed as a consequence of Theorem 4.10 by L. E. Dubins and D. A. Freedman in their 1981 paper, in which the original proof of the latter theorem appeared. A much shorter proof of Theorem 4.10 appeared in Gale and Sotomayor's (1985a) paper using Lemma 3.5, stated by Hwang (n.d.), who observed that it could be employed in such a proof. The observation that these theorems did not depend on the assumption of strict preferences was made in Roth (1984b). Theorem 4.11 is from Demange, Gale, and Sotomayor (1987).

Related strategic results for a one-sided "housing market" model are found in Roth (1982b) and Bird (1984). As remarked earlier, that the "sports draft" procedure discussed in Example 4.3 makes it a dominant strategy for each agent to state his true preferences depends on the fact that we are dealing with one-to-one matching. In actual sports drafts, in which teams draft one player per round in each of several rounds, it is not a dominant strategy for each team to draft players in the order in which it prefers them, and at equilibrium the results may not be Pareto optimal. This is discussed by Brams and Straffin (1979).

The proofs of Theorems 4.15, 4.17, and 4.21, as well as Example 4.18 are due to Gale and Sotomayor. Theorems 4.16 and 4.20 are due to Roth (1984b, 1982a, respectively), using proofs based on the deferred acceptance procedure (as in the alternate proof sketched here for Theorem 4.16). The proofs of all these results presented here follow Gale and Sotomayor (1985b). The general notion of strategic equilibrium is due to Nash (1951). An introductory treatment of games that pays particular attention to dominating strategies (and is written by a scholar whose own work has considerably enhanced our understanding of these matters) is found in Moulin (1986). Roth and Vande Vate (1989) discuss results related to Theorem 4.16 for the case of a matching mechanism that randomly chooses a stable matching with respect to the stated preferences.

The model and all the results of Section 4.5, concerning matching games of incomplete information about others' preferences, come from Roth (1989a). The general formulation of games of incomplete information originates with Harsanyi (1967, 1968a, b). The "revelation principle" (Section 4.5.1) is a widely used proof technique for games of incomplete information: See Myerson (1985) for a good introductory treatment of this and related matters by an author who has contributed as much as anyone to our understanding of this kind of tool.

Another class of incomplete information models, different from that of Section 4.5 but also relevant to the general kind of problem considered here, is found in what is called the "job search" literature. These models address the question of how firms and workers become informed about one another when search is costly. So in these models agents can be viewed as not knowing their own preferences with certainty, as these preferences are determined by attributes of the other players about which they have incomplete information. Some representative papers from this literature are Mortensen (1982) and Diamond and Maskin (1979, 1982).

In closing we should mention that many of the results of this chapter are in the "style" of the branch of economic theory known as social choice theory. Indeed, the seminal "impossibility" theorem is Arrow's 1951 demonstration that certain kinds of "social choice functions" are impossible. Gibbard (1973) and Satterthwaite (1975) showed how this result was closely related to the impossibility of certain kinds of general procedures that make it a dominant strategy for agents to reveal their true preferences. Peleg (1978) suggested that strategic and strong equilibrium misstatements of preferences should therefore be of interest. Some idea of the literature concerning what kinds of results can be "implemented" in terms of which kinds of strategic equilibria can be gained from Dasgupta, Hammond, and Maskin (1979), and from Green and Laffont (1979) and Peleg (1984), who consider particular domains on which progress can be made. Some results in this connection for a model of "personal incomplete information" that includes the kind of incomplete information modeled in Section 4.5 are found in Kalai and Samet (1985).

PART II

Many-to-one matching: models in which firms may employ many workers

Chapters 5 and 6 will explore models of many-to-one matching. In terms of the economic phenomena that motivate the models in this book, many-to-one matching in two-sided markets is perhaps the most typical case, where one side of the market consists of institutions and the other side of individuals. Thus colleges admit many students, firms hire many workers, and hospitals employ many interns, all at the same time. But students typically attend only one college (at least at any given time), and so forth.

A central issue in formulating a model of many-to-one matching will be how to model the preferences of the institutions, since these involve comparisons of different *groups* of students, workers, and so on. No comparable question arose in the marriage model, where preferences over individuals were sufficient to determine preferences over matchings. We will see that there are both important differences and striking similarities between one-to-one and many-to-one matching. The results presented in Chapter 5 will also complete the explanation begun in Section 1.1 of the phenomena described there concerning the labor market for medical interns. And some further empirical observations in related markets will briefly be described.

The chief difference between the model explored in Chapter 5 and the two models presented in Chapter 6 is in the approach they take to modeling the preferences of the institutions. In Chapter 5 we consider preferences over groups of students, interns, and the like, to be derived from preferences over individuals, whereas in Chapter 6 we look directly at preferences over groups and consider a more general family of preferences. Another difference is that in Chapter 5 we model monetary payments (salaries, tuition, etc.) only implicitly as one of the features that determine preferences, and in Chapter 6 we model wages explicitly.

CHAPTER 5

The college admissions model and the labor market for medical interns

5.1 The formal model

There are clear similarities between the hospital intern market and the simple model of a marriage market studied in the previous chapters. There are two kinds of agents, hospitals and medical students, and the function of the market is to match them. (Strictly speaking, we should speak of hospital programs rather than hospitals, because different internship programs within a hospital are separately administered, and students apply to specific programs.) Because interns' salaries are part of the job description of each position, and not negotiated as part of the agreement between each hospital and intern, salaries will not play an explicit role in our model, but will simply be one of the factors that determine the preferences that students have over the hospitals. Similarly, we will assume that hospitals have preferences over students – that is, they are able to rank order the students who have applied to them for positions, as they are asked to do by the National Resident Matching Program. The major difference from the marriage problem is that each hospital program may employ more than one student, although each student can take only one position. (All the positions offered by a given hospital program are identical, since hospitals offering different kinds of positions must divide them into different programs.) The rules of the market are that any student and hospital may sign an employment contract with each other if they both agree, any hospital may choose to keep one or more of its positions unfilled, and any student may remain unmatched if he or she wishes (and seek employment later in a secondary market).

Of course, there are many other matching problems that fit this general description in which the agents on one side of the market are individuals and the agents on the other side are institutions (e.g., students and colleges or firms and workers). The formal model presented here is a reformulation of what is most often referred to in the theoretical literature

125

as the "college admissions" model. We will adopt that terminology here, and in most of this chapter refer to the institutions as colleges. The exception will be in Section 5.4, which presents results directly motivated by the study of the labor market for medical interns. In that section the institutions will be referred to as hospitals.

So the first elements of our formal model are two finite and disjoint sets, $C = \{C_1, ..., C_n\}$ and $S = \{s_1, ..., s_m\}$, of colleges and students, respectively. Each student has preferences over the colleges, and each college has preferences over the students. As in the marriage model, we will assume these preferences are complete and transitive, so they may be represented by ordered lists, with $P(C) = s_1, s_2, C, s_3, ...$ denoting that college C prefers to enroll s_1 rather than s_2, that it prefers to enroll either one of them rather than leave a position unfilled, and that all other students are unacceptable, in the sense that it would be preferable to leave a position unfilled rather than filling it with, say, student s_3. Similarly, $P(s) = C_2, C_1, C_3, s, ...$ represents the preferences of student s precisely as in the marriage problem, indicating for example that the only positions the student would accept are those offered by C_2, C_1, and C_3, in that order. We will write $C_i >_s C_j$ to indicate that student s prefers C_i to C_j, and $C_i \geq_s C_j$ to indicate that s likes C_i at least as well as C_j; that is, that he or she either prefers C_i to C_j or is indifferent between them. Similarly, $s_i >_C s_j$ and $s_i \geq_C s_j$ represent college C's preferences $P(C)$ over individual students. College C is *acceptable* to student s if $C \geq_s s$, and student s is acceptable to college C if $s \geq_C C$. As in the marriage problem, we will abbreviate preference lists to include just the acceptable alternatives.

The first difference from the marriage model is that for each college C, there is a positive integer q_C called the *quota* of college C, which indicates the number of positions it has to offer, that is, the maximum number of positions it may fill. (That all q_C positions are identical is reflected in the fact that students' preferences are over colleges – they do not distinguish between positions.) When we denote a particular college by C_i, its quota will be denoted q_i.

An outcome of the college admissions model is a matching of students to colleges, such that each student is matched to at most one college, and each college is matched to at most its quota of students. A student who is not matched to any college will be "self-matched" as in the marriage problem, and a college that has some number of unfilled positions will be matched to itself in each of those positions. A matching is bilateral, in the sense that a student is enrolled at a given college if and only if the college enrolls that student.

To give a formal definition, we first define for any set X an *unordered family of elements of X* to be a collection of elements, *not necessarily* *distinct,* in which the order is immaterial. So a given element of X may appear more than once in an unordered family of elements of X, which is what distinguishes an unordered family from a subset of X. We can now formally define a matching as follows.

Definition 5.1. *A matching μ is a function from the set $C \cup S$ into the set of unordered families of elements of $C \cup S$ such that:*

1. $|\mu(s)| = 1$ *for every student s and $\mu(s) = s$ if $\mu(s) \notin C$;*
2. $|\mu(C)| = q_C$ *for every college C, and if the number of students in $\mu(C)$, say r, is less than q_C, then $\mu(C)$ contains $q_C - r$ copies of C;*
3. $\mu(s) = C$ *if and only if s is in $\mu(C)$.*

So $\mu(s_1) = C$ denotes that student s_1 is enrolled at college C at the matching μ, and $\mu(C) = \{s_1, s_3, C, C\}$ denotes that college C, with quota $q_C = 4$, enrolls students s_1 and s_3 and has two positions unfilled. (We continue to write the elements of $\mu(C)$ between brackets normally reserved for sets.)

We will represent matchings graphically, for example,

$$\mu_1 \quad \begin{array}{cccc} C_1 & C_2 & (s_4) \\ s_1\ s_3\ C_1 & s_2 & s_4 \end{array}$$

represents a matching at which college C_1, with quota $q_{C_1} = 3$, is matched with two students, s_1 and s_3, college C_2 with a quota of one is matched with one student, and student s_4 is unmatched.

5.1.1 Preferences over matchings

At this point in our description of the marriage model, we had only to say that each agent's preferences over alternative matchings correspond exactly to his or her preferences over his or her own assignments at the two matchings. We can now say the same thing about students, since at each matching a student is either unmatched or matched to a college, and we have already described students' preferences over colleges. But, even though we have described colleges' preferences over students, each college with a quota greater than one must be able to compare *groups* of students in order to compare alternative matchings, and we have yet to describe the preferences of colleges over groups of students. (Until we have described colleges' preferences over matchings, our model will not be a well-defined game.)

The simplest assumption connecting colleges' preferences over groups of students to their preferences over individual students is one insuring that, for example, if $\mu(C)$ assigns college C its third and fourth choice students, and $\mu'(C)$ assigns it its second and fourth choice students, then college C prefers $\mu'(C)$ to $\mu(C)$. Specifically, let $P^{\#}(C)$ denote the preference relation of college C over all assignments $\mu(C)$ it could receive at some matching μ of the college admissions problem. A college C's preferences $P^{\#}(C)$ will be called *responsive* to its preferences $P(C)$ over individual students if, for any two assignments that differ in only one student, it prefers the assignment containing the more preferred student (and is indifferent between them if it is indifferent between the students). We can state the definition formally, as follows.

Definition 5.2.   *The preference relation $P^{\#}(C)$ over sets of students is responsive (to the preferences $P(C)$ over individual students) if, whenever $\mu'(C) = \mu(C) \cup \{s_k\} \setminus \{\sigma\}$ for σ in $\mu(C)$ and s_k not in $\mu(C)$, then C prefers $\mu'(C)$ to $\mu(C)$ (under $P^{\#}(C)$) if and only if C prefers s_k to σ (under $P(C)$).*

We will write $\mu'(C) >_C \mu(C)$ to indicate that college C prefers $\mu'(C)$ to $\mu(C)$ according to its preferences $P^{\#}(C)$, and $\mu'(C) \geq_C \mu(C)$ to indicate that C likes $\mu'(C)$ at least as well as $\mu(C)$, where the fact that $\mu'(C)$ and $\mu(C)$ are not singletons will make clear that we are dealing with the preferences $P^{\#}(C)$, as distinct from statements about C's preferences over individual students. Note that C may be indifferent between distinct assignments $\mu(C)$ and $\mu'(C)$ even if C has strict preferences over individual students.

Note also that different responsive preference orderings $P^{\#}(C)$ exist for any preference $P(C)$, since, for example, responsiveness does not specify whether a college with a quota of two prefers to be assigned its first and fourth choice students instead of its second and third choice students. However the preference ordering $P(C)$ over individual students can be derived from $P^{\#}(C)$ by considering a college C's preferences over assignments $\mu(C)$ containing no more than a single student (and $q_C - 1$ copies of C). The assumption that colleges have responsive preferences is essentially no more than the assumption that their preferences for sets of students are related to their ranking of individual students in a natural way. (In Chapter 6 we will consider more complicated kinds of preferences, which arise when preferences are such that firms cannot be said to have rankings of individual employees.)

We will henceforth assume that colleges have preferences over groups of students that are responsive to their preferences over individual students as well as being complete and transitive, and that each agent's preferences over alternative matchings correspond exactly to his or her (its) preferences over his or her (its) own assignments at the two matchings.

As in the marriage model, some of the results that follow will depend on the assumption that agents have strict preferences. Surprisingly, we will only need to assume that colleges have strict preferences over individuals: It will not be necessary to assume that they have strict preferences over groups of students. The reasons for this will not become completely clear until Theorem 5.26, which says that when colleges have strict preferences over individuals, then they are not indifferent between any groups of students assigned to them at stable matchings, even though they may be indifferent between other groups of students.

5.2   Stability and group stability

A matching μ is *individually irrational* if $\mu(s) = C$ for some student s and college C such that either the student is unacceptable to the college or the college is unacceptable to the student. Such a matching will also be said to be *blocked* by the unhappy agent. As in the marriage problem, this terminology reflects that the rules of the market allow every agent to withhold his or her consent from such a match.

Similarly, a collge C and student s will be said together to block a matching μ if they are not matched to one another at μ, but would both prefer to be matched to one another than to (one of) their present assignments. That is, μ is *blocked by the college-student pair* (C, s) if $\mu(s) \neq C$ and if $C >_s \mu(s)$ and $s >_C \sigma$ for some σ in $\mu(C)$. (Note that σ may equal either some student s' in $\mu(C)$, or, if one or more of college C's positions is unfilled at $\mu(C)$, σ may equal C.) It should be clear that matchings blocked in this way by an individual or by a pair of agents are unstable in the sense discussed for the marriage model, since there are agents with both the incentive (because preferences are responsive) and the power to disrupt such matchings. So, as in the marriage model, we now define stable matchings – although we will immediately have to turn to the question of whether the set of stable matchings defined in this way can serve the same role as it did in the marriage model.

Definition 5.3. *A matching μ is **stable** if it is not blocked by any individual agent or any college-student pair.*

On the face of it, it isn't obvious that this definition, which is a definition of *pairwise* stability, will be adequate, since we now might need to consider coalitions consisting of a college and several students (all of whom might be able to enroll simultaneously at the college), or even coalitions consisting of multiple colleges and students. We shall consider these larger coalitions now, and see that when preferences are responsive, nothing is lost by concentrating on simple college–student pairs.

We will call a matching μ *group unstable*, or say it is *blocked by a coalition A*, if there exists another matching μ' and a coalition A, which might consist of multiple students and/or colleges, such that for all students s in A, and for all colleges C in A,

1. $\mu'(s) \in A$ (i.e., every student in A who is matched by μ' is matched to a college in A);
2. $\mu'(s) >_s \mu(s)$ (i.e., every student in A prefers his or her new match to his or her old one);
3. $\sigma \in \mu'(C)$ implies $\sigma \in A \cup \mu(C)$ (i.e., every college in A is matched at μ' to new students only from A, although it may continue to be matched with some of its "old" students from $\mu(C)$); and
4. $\mu'(C) >_C \mu(C)$ (i.e., every college in A prefers its new set of students to its old one).

That is, μ is blocked by some coalition A of colleges and students if, by matching among themselves, the students and colleges in A could all get an assignment preferable to μ.

Definition 5.4.   *A group stable matching is one that is not blocked by any coalition.*

We will now see that (when preferences are responsive) this definition of group stability is entirely equivalent to our definition of (pairwise) stability for the college admissions model.

Lemma 5.5.   *A matching is group stable if and only if it is (pairwise) stable.*

Proof: If μ is unstable via an individual student or college, or via a student–college pair, then it is clearly group unstable via the coalition consisting of the same singleton or pair.

In the other direction, if μ is blocked via coalition A and outcome μ', let C be in A. Then the fact that $\mu'(C) >_C \mu(C)$ implies that there exists a student s in $\mu'(C) - \mu(C)$ and a σ in $\mu(C) - \mu'(C)$ such that $s >_C \sigma$. (Otherwise, $\sigma \geq_C s$ for all σ in $\mu(C) - \mu'(C)$ and s in $\mu'(C) - \mu(C)$, and this would imply $\mu(C) \geq_C \mu'(C)$, by repeated application of the fact that preferences are responsive and transitive.) So s is in A and s prefers C to $\mu(s)$, so μ is unstable via s and C.

Lemma 5.5 has an obvious relationship to Theorem 3.3 of Section 3.1, since both say that the instabilities that can arise from coalitions of any size can be identified by examining only small coalitions. As we shall see in Section 5.7, this does not carry over into precisely the same identity that we found for the marriage model between the set of stable matchings and the core.

5.3   The connection between the college admissions model and the marriage model

The importance of Lemma 5.5 for the college admissions model goes beyond the fact that it allows us to concentrate on small coalitions. It says that stable and group stable matchings can be identified using only the preferences P over *individuals* – that is, without knowing the preferences $P^{\#}(C)$ that each college has over *groups* of students. (This would not be true if preferences were not responsive; see e.g. Example 6.6.) So the set of stable (or group stable) matchings will not be sensitive to changes in the preferences $P^{\#}(C)$ (so long as these preferences remain responsive to the preferences $P(C)$).

This suggests that the college admissions model may be very similar indeed to the marriage model, and that many of the results obtained for the marriage model will generalize immediately to the college admissions model. This turns out to be an issue about which there has been confusion in the literature. We turn now to consider it.

5.3.1   A related marriage market

Consider a particular college admissions problem, with colleges $C = \{C_1, \ldots, C_n\}$ having quotas $q_1, \ldots, q_n$, and students $S = \{s_1, \ldots, s_m\}$. The preferences of students and colleges over individuals are given by $P = \{P(C_1), \ldots, P(C_n); P(s_1), \ldots, P(s_m)\}$.

We can consider a related marriage market, in which each college C with quota q_C is broken into q_C "pieces" of itself, so that in the related market, the agents will be students and college positions, each having a quota of one. That is, we replace college C by q_C positions of C denoted by $c_1, c_2, \ldots, c_{q_C}$. Each of these positions has preferences over individuals that are identical with those of C. Since each position c_i has a quota of one, we do not need to consider its preferences over groups of students.

We have some leeway in describing the preferences of the students, who are in fact indifferent between the different positions at each college that are now each represented as a separate agent. In order not to complicate the exposition about results for which the assumption of strict preferences is important, we will assume that each student's preference list is modified by replacing C, wherever it appears on his or her list, by the string $c_1, c_2, \ldots, c_{q_C}$, in that order. That is, if student s in the original college admissions problem preferred C to C', then student s in the marriage market we are constructing prefers all positions of C to all positions of C', and we assume for convenience that s strictly prefers c_1, say, to all the other positions of C, and so forth.

If the preferences over individuals are strict, there is a natural one-to-one correspondence between matchings in the original college admissions problem and matchings in the marriage market derived from it in this way. That is, a matching μ of the college admissions problem, which matches a college C with the students in $\mu(C)$, corresponds to the matching μ' in the related marriage market in which the students in $\mu(C)$ are matched, in the order that they occur in the preferences $P(C)$, with the ordered positions of C that appear in the related marriage market. (That is, if s is C's most preferred student in $\mu(C)$, then $\mu'(s) = c_1$, and so forth.) Furthermore, this one-to-one correspondence preserves the stability of the matching. (Note that since all the agents $c_1, c_2, ..., c_{q_C}$ corresponding to a college C have the same preferences, in order for the matching μ' corresponding to a stable matching μ in the college admissions problem to be a stable matching in the marriage market, it *must* be that the most preferred student in $\mu(C)$ is married to c_1, the second most preferred to c_2, and so forth.) If the preferences over individuals are not strict (specifically, if colleges are indifferent between some students), there can be more than one matching in the related marriage market corresponding to a given matching of the college admissions problem, since the position of college $\mu(s) = C$ to which a student s is matched in the related marriage market need not be uniquely determined.

We have the following lemma, whose proof we leave to the reader.

Lemma 5.6. *A matching of the college admissions problem is stable if and only if the corresponding matchings of the related marriage market are stable.*

Thus some of the theorems we proved for the marriage model, such as Theorem 2.8 that says that the set of stable matchings is nonempty for every marriage market, will immediately generalize to the case of the college admissions model, via Lemma 5.6. (Of course, this result depends on colleges having responsive preferences, since if not, the set of stable matchings might be empty as in Example 2.7.) In the following sections we will speak of several other theorems that generalize via Lemma 5.6, which establishes an important relationship between the two kinds of markets.

Another important relationship is that the class of marriage markets is the special case of college admissions problems in which all colleges have a quota of one. This allows the impossibility theorem (Theorem 4.4) to be immediately generalized to the college admissions model. That is, the impossibility theorem says that there are no stable matching mechanisms (defined on the class of all marriage markets) that make it a dominant strategy for all agents to state their true preferences. If we now define stable matching mechanisms on the class of college admissions problems (so a stable matching mechanism is a function defined on all college admissions problems (S, C, P) whose output is a stable matching with respect to P) it follows immediately that no such mechanism can exist that makes it a dominant strategy for all agents to state their true preferences in all college admissions problems (since if there were, then we could just apply it to the special class of marriage problems, in contradiction to the impossibility theorem).

That is to say, a negative result like the impossibility theorem, which says it is impossible to design a mechanism that performs in a certain way for every marriage market, generalizes to college admissions problems by the observation that marriage markets *are* college admissions problems. And certain kinds of positive results, such as the existence of stable matchings, generalize via Lemma 5.6, which establishes a connection between the stable matchings in the two kinds of markets. We will nevertheless see that college admissions problems have some important differences from simple marriage markets. We turn next to a preliminary discussion of what kinds of results might not generalize from marriage markets.

5.3.2   *Limitations to the relationship with the marriage market*

The differences between college admissions problems and marriage markets will appear in different ways depending on whether we are looking at results concerning the structure of the set of stable matchings, or results concerning strategic decisions made by the agents.

Lemma 5.6 establishes the connection between stable matchings of college admissions problems and marriage markets. However, since stable matchings can be identified without regard to the preferences of colleges over groups of students, this result does not permit us to directly conclude anything about the preferences of colleges for different (stable or unstable) matchings. Thus results like Theorems 2.12, 2.13, and 2.16 concerning optimal stable matchings and the comparison of different matchings, as well as results like Theorems 2.24, 2.25, and 2.26, which compare stable matchings in different markets, will have to be considered again. So will Theorem 2.27, which compares stable and unstable matchings. We will see that most of these theorems do in fact generalize to the college admissions model, sometimes with additional power. But care must be taken, since some results that hold for marriage markets, such as the conclusions of Theorem 2.27, turn out not to hold for college admissions problems. (And, once again, if preferences are not responsive it is not in general possible even to define stable matchings without considering preferences over groups of students (see Example 6.6), and the set of stable matchings may be empty, as in Example 2.7.)

When we consider strategic issues, we will see that there are quite substantial differences between the two kinds of markets. This is because any college with a quota greater than one resembles something like a coalition, rather than an individual, in the related marriage market. This will allow us to strengthen the conclusions of the impossibility theorem (see Theorem 5.14), but will force us to weaken a number of our other conclusions about strategic behavior in marriage markets (compare Theorem 4.7 with Theorem 5.14, Theorems 4.10 and 4.11 with Corollary 5.15, and Theorem 4.16 with Corollary 5.17).

We begin by establishing the connection suggested in Section 1.1 between the college admissions model and the history and organization of the labor market for American medical interns.

5.4   The labor market for medical interns

We already know via Lemma 5.6 that the set of stable matchings for the hospital–intern market is never empty. It is also easy to see how the deferred acceptance algorithm described in Chapter 2 could be modified to give us a direct proof – if the hospitals are doing the proposing, then each hospital makes as many proposals at each stage as needed to bring it up to its quota of "engagements," whereas if the students are doing the proposing, then each hospital starts to reject acceptable students only when its quota is full. Modified in this way, the deferred acceptance algorithm is easily seen to be a stable matching mechanism for the hospital–intern market – that is, for any stated preferences P, it produces a stable matching with respect to P.

However since our purpose in this section is to try to explain some of the events described in Section 1.1, it will not be sufficient for us merely to prove that stable matchings exist, or to study the properties of an arbitrary stable matching mechanism. Instead, we need to study the particular algorithm adopted in this market in 1951, and see what properties it has. After describing it, we will see that it too is a stable matching mechanism for the hospital–intern market.

5.4.1   *The NIMP algorithm*

The following description is given in Roth (1984a).

The NIMP algorithm...works as follows. Each hospital program rank orders the students who have applied to it (marking "X" any students who are unacceptable) and each student rank orders the hospital programs to which he has applied (similarly indicating any which are unacceptable). These lists are mailed to the central clearinghouse, where they are edited by removing from each hospital program's rank-order list any student who has marked that program as unacceptable, and by removing from each student's list any hospital which has indicated he is unacceptable.... The edited lists are thus rank orderings of acceptable alternatives.

These lists are entered into what may be thought of as a list-processing algorithm consisting of a *matching phase* and a *tentative-assignment-and-update* phase. The first step of the matching phase (the *1:1* step) checks to see if there are any students and hospital programs which are top-ranked in one another's ranking. (If a hospital H_i has a quota of q_i then the q_i highest students in its ranking are top-ranked.) If no such matches are found, the matching phase proceeds to the *2:1* step, at which the second ranked hospital program on each student's ranking is compared with the top-ranked students on that hospital's ranking. At any step when no matches are found, the algorithm proceeds to the next step, so the generic *k:1* step of the matching phase seeks to find student-hospital pairs such that the student is top-ranked on the hospital's ranking and the hospital is kth ranked by the student. At any step where such matches are found, the algorithm proceeds to the tentative-assignment-and-update phase.

When the algorithm enters the tentative-assignment-and-update phase from the $k:1$ step of the matching phase, the $k:1$ matches are tentatively made; i.e., each student who is a top-ranked choice of his kth choice hospital is tentatively assigned to that hospital. The rankings of the students and hospitals are then updated in the following way. Any hospital which a student s_j ranks lower than his tentative assignment is deleted from his ranking (so the updated ranking of a student s_j tentatively assigned to his kth choice now lists only his first k choices) and student s_j is deleted from the ranking of any hospital which was deleted from s_j's ranking (so the updated rankings of each hospital now include only those applicants who haven't yet been tentatively assigned to a hospital they prefer). Note that, if one of a hospital's top-ranked candidates is deleted from its ranking, then a lower-ranked choice moves into the top-ranked category, since the hospital's updated ranking has fewer students, but the same quota, as its original ranking. When the rankings have been updated in this way, the algorithm returns to the start of the matching phase, which examines the updated rankings for new matches. Any new tentative matches found in the matching phase replace prior tentative matches involving the same student. (Note that new tentative matches can only improve a student's tentative assignment, since all lower ranked hospitals have been deleted from his ranking.) The algorithm terminates when no new tentative matches are found, at which point tentative matches become final. That is, the algorithm matches students with the hospitals to which they are tentatively matched when the algorithm terminates. Any student or hospital position which was not tentatively matched during the algorithm is left unassigned, and must make subsequent arrangements by directly negotiating with other unmatched students or hospitals.

Note that the procedure does not allow hospitals or students to express indifference between alternatives, so the submitted rank orderings

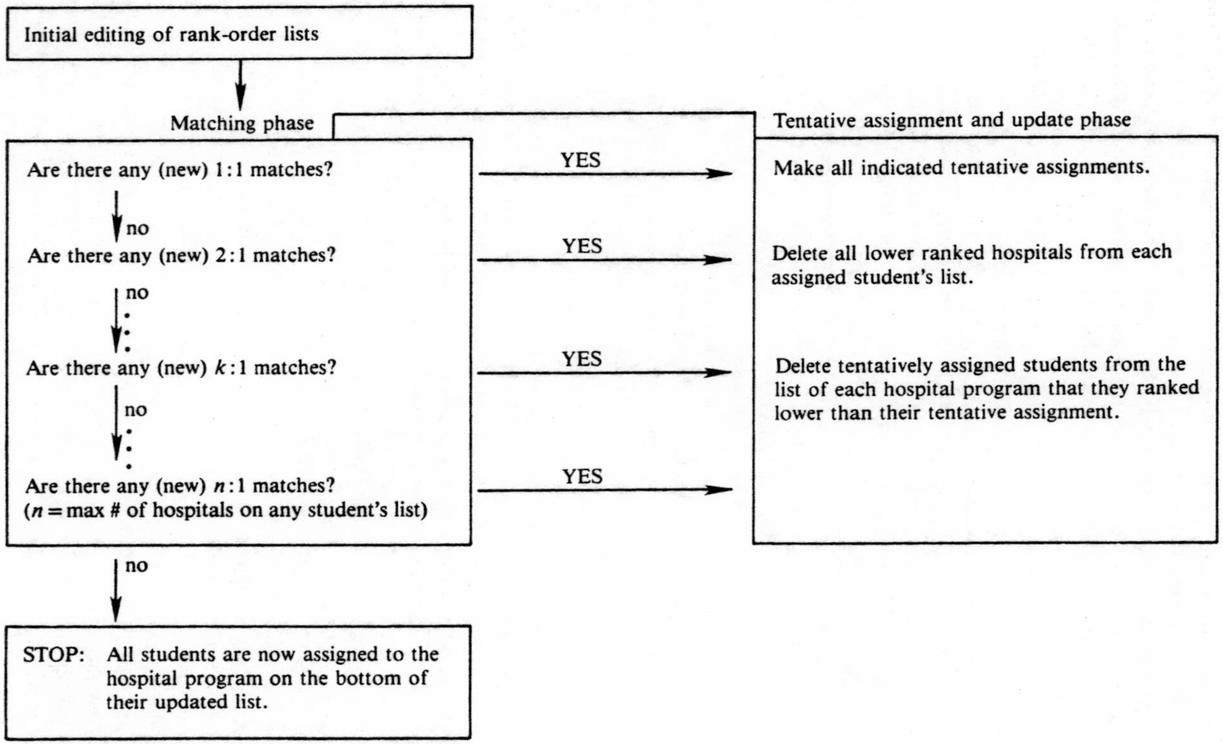

Figure 5.1. The NIMP algorithm.

Initial editing of rank-order lists

Matching phase

Are there any (new) 1:1 matches?

Are there any (new) 2:1 matches?

Are there any (new) k:1 matches?

Are there any (new) n:1 matches?
(n = max # of hospitals on any student's list)

no

no

no

no

YES

YES

YES

YES

Tentative assignment and update phase

Make all indicated tentative assignments.

Delete all lower ranked hospitals from each assigned student's list.

Delete tentatively assigned students from the list of each hospital program that they ranked lower than their tentative assignment.

STOP: All students are now assigned to the hospital program on the bottom of their updated list.

are strict preference lists. See Figure 5.1 (reproduced from Roth, 1984a) for a schematic of the algorithm.

To see the critical role played by the initial editing of the submitted preference lists, consider an example with three students and two hospitals, each with a quota of one, with submitted preference lists as follows (only acceptable alternatives are listed).

$$P(H_1) = s_1, s_2, s_3; \qquad P(H_2) = s_1, s_2, s_3$$

$$P(s_1) = H_1, H_2; \qquad P(s_2) = H_1; \qquad P(s_3) = H_1, H_2$$

The unique stable matching in this example is

$$\mu = \frac{H_1 \; H_2 \; (s_2)}{s_1 \; s_3 \; s_2},$$

but if the matching and tentative assignment phases of the NIMP algorithm are conducted on the unedited preferences, the unstable matching

$$\mu' = \frac{H_1 \; H_2 \; (s_2) \; (s_3)}{s_1 \; (H_2) \; s_2 \; s_3}$$

results, since no further $k:1$ matches are found after s_1 is matched to H_1. This is because the remaining hospital with an unfilled quota, H_2, has s_2 top-ranked, but H_2 isn't listed by s_2. When the initial editing step of the algorithm is first applied, however, H_2's edited preference list becomes $P(H_2) = s_1, s_3$ and the algorithm produces the stable matching μ.

Note that the NIMP algorithm is thus substantially different from the deferred acceptance algorithm modified to allow quotas greater than one, since, for example, the initial editing of preferences would make no difference in the deferred acceptance algorithm. Surprisingly, these two different algorithms are equivalent as stable matching mechanisms, as we will now see.

5.4.2 The NIMP algorithm is a stable matching mechanism

This section shows that the NIMP algorithm is a stable matching mechanism that produces, for any stated preferences, the hospital-optimal stable matching.

The proofs of Theorems 5.7 and 5.8 closely follow our treatment of the deferred acceptance algorithm, and together amount to a formal proof that the NIMP algorithm is equivalent as a matching mechanism to the deferred acceptance algorithm (with hospitals proposing, and taking into account that they can propose to more than one student).

Theorem 5.7 (Roth).   *The NIMP algorithm is a stable matching mechanism, that is, it produces a stable matching with respect to any stated preferences.*

Proof: When the algorithm terminates, each hospital H_i is matched with the top q_i choices (i.e., the top-ranked choices) on its final updated rank-order list. (This follows since the algorithm doesn't terminate while tentative $k:1$ matches can still be found.) This assignment is stable, since any student s_j who some hospital H_i originally ranked higher than one of its final assignees was deleted from H_i's ranking when s_j was given a tentative assignment higher in his or her ranking than H_i. Hence the final assignment gives s_j a position he or she ranked higher than H_i. So the final matching is not unstable with respect to any such H_i and s_j.

Theorem 5.7 thus provides the first link in the explanation outlined in Section 1.1 for why the NIMP algorithm was able to achieve such high rates of voluntary participation, in contrast to the chaotic "recontracting" that characterized the years 1945–51, before the centralized procedure was introduced.

For a given vector P of preferences over individuals we continue to say (as in the marriage model) that hospital H and student s are *achievable* for one another if there is some stable matching at which they are matched. For each H_i with quota q_i, let a_i be the number of achievable students, and define $k_i = \min\{q_i, a_i\}$. Then for any vector P of rank-order lists submitted by the hospitals and students (which are strict preference lists over individuals, since no indifference may be indicated on the lists), the NIMP algorithm produces a matching that is hospital-optimal in the following strong sense.

Theorem 5.8 (Roth).   *For any submitted lists of (strict) preferences over individuals, the NIMP algorithm produces a matching that gives each hospital H_i its k_i highest ranked achievable students.*

Proof: It will be sufficient to show that no achievable student is ever deleted from a hospital's rank-order list, so the final assignment gives each hospital its top q_i achievable students, if it has that many achievable students, or all its achievable students if $a_i < q_i$. This can be seen by induction. Suppose that, up to the rth iteration of the algorithm, no student has been deleted from the ranking of a hospital for whom he or she is achievable, and that on the $(r+1)$st iteration student s_j is tentatively matched with hospital H_i, and deleted from the ranking of hospital H_k. Then any assignment that matches s_j with H_k, and assigns achievable matches to H_i, is unstable since s_j ranked H_i higher than H_k and H_i ranked s_j higher than one of its assignees. (This follows since s_j was top-ranked by H_i at the end of the rth iteration, when no achievable students had yet been deleted from H_i's rank-order list.) So s_j is not achievable for H_k.

Note that the assumption that hospitals have responsive preferences implies they prefer the matching resulting from the NIMP algorithm to any other stable matching with respect to the submitted preferences lists – that is, the matching is the (unique) hospital-optimal stable matching. Note also that the consequences of optimality are stronger here than in the marriage model. Whereas the men at the M-optimal stable matching of a marriage problem each receive their most preferred achievable mate, here each hospital receives its k_i most preferred achievable students. We can state the following corollary.

Corollary 5.9. *When preferences are strict, there exists a hospital-optimal stable matching that every hospital likes as well as any other stable matching, and a student-optimal stable matching that every student likes as well as any other stable matching.*

The existence of the student-optimal stable matching when preferences are strict follows as in Chapter 2.

Whereas the previous results have resembled those for marriage markets, the following theorem is quite different from the related result for the marriage market.

Theorem 5.10 (Roth). *When the preferences over individuals are strict, the student-optimal stable matching is weakly Pareto optimal for the students, but the hospital-optimal stable matching need not be even weakly Pareto optimal for the hospitals.*

Proof: That no matching is preferred by every student to the student-optimal stable matching follows from the similar result for the marriage problem (Theorem 2.27) and Lemma 5.6.

The proof that the analogous result does not hold for the hospital side of the market will be by means of an example. Consider the problem consisting of three hospitals $H = \{H_1, H_2, H_3\}$ and four students $S = \{s_1, s_2, s_3, s_4\}$. Hospital H_1 has a quota of $q_1 = 2$, and both other hospitals have a quota of one. The true preferences P are given by:

$$P(s_1) = H_3, H_1, H_2 \qquad P(H_1) = s_1, s_2, s_3, s_4$$
$$P(s_2) = H_2, H_1, H_3 \qquad P(H_2) = s_1, s_2, s_3, s_4$$
$$P(s_3) = H_1, H_3, H_2 \qquad P(H_3) = s_3, s_1, s_2, s_4$$
$$P(s_4) = H_1, H_2, H_3.$$

It is straightforward to verify that the hospital-optimal (and only) stable matching is the matching μ_H given by $\mu_H(H_1) = \{s_3, s_4\}$, $\mu_H(H_2) = \{s_2\}$, and $\mu_H(H_3) = \{s_1\}$.

Consider now the matching μ' such that $\mu'(H_1) = \{s_2, s_4\}$, $\mu'(H_2) = \{s_1\}$, and $\mu'(H_3) = \{s_3\}$. The matching μ' gives hospitals H_2 and H_3 each their first choice student, so they both prefer it to μ_H. Since Hospital H_1 has responsive preferences, it strictly prefers μ' to μ_H. Thus every hospital prefers μ' to μ_H.

To understand why the hospital–intern market is different from the marriage market in this respect, recall Example 2.31, which showed that the optimal stable matching for one side of the marriage market need only be weakly Pareto optimal for that side of the market. That is, although there is no matching that all men, say, prefer to the man-optimal stable matching in a marriage market, in the example all men but one prefer some other matching, and the one remaining man received the same spouse at both matchings. Now, in the marriage market related to a hospital–intern market, it is the hospital positions that play the role of the agents on the hospital side of the market. So Theorem 2.27 and Lemma 5.6 tell us that there exists no matching that gives every hospital a more preferred student *in every position* than it gets at the hospital-optimal stable matching. But of course, as we have just seen, this does not imply that the hospitals don't all prefer some other matching.

5.4.3 *Married couples*

In Section 1.1 we noted that, at least as early as 1973, significant numbers of married couples declined to take part in the NIMP procedure, or to accept jobs assigned to them by that procedure. In view of the preceding discussion, this should make us suspect that something about the presence of couples introduced instabilities into the market. In fact, the NIMP program included a specific procedure for handling couples that will make it fairly clear how these instabilities arose (and why they were so prevalent), at least until the mid-1980s, when the procedure for married couples was modified. We will also see, however, that the problem goes deeper than the particular procedure in use at any time, because the introduction of married couples into the market may make all matchings unstable.

Briefly, the situation prior to the mid-1980s was this. Couples graduating from medical school at the same time and wishing to obtain two positions in the same community had two options. One option was to stay outside of the NIMP program and negotiate directly with hospital programs. (Hospitals that negotiated directly with a member of such a couple could still seek to fill their remaining positions through the centralized matching program.) Alternatively, they could (after being certified by the dean of their medical school as a legitimate couple) enter the NIMP program together as a couple to be matched by a special "couples algorithm."

This couples algorithm can be described roughly as follows. The couple was required to specify one of its members as the "leading member," and to submit a rank ordering of positions for each member of the couple; that is, a couple submitted two preference lists, one for each member. The leading member of the couple was then matched to a position in the usual way, the preference list of the other member of the couple was edited to remove distant positions, and the second member was then matched if possible to a position in the same vicinity as the leading member.

It is easy to see why instabilities would often result. Consider a couple $\{s_1, s_2\}$ whose first choice is to have two particular jobs in Boston, and whose second choice is to have two particular jobs in New York. (If s_1 and s_2 are interested in different medical specialties, or have different medical school records, it may or may not be feasible for them to contemplate getting two jobs in a single hospital program. Typically they will have to deal with two separate hospital programs.) Under the couples algorithm, the designated leading member might be matched to his or her first choice job in Boston, while the other member might be matched to some relatively undesirable job in Boston. If s_1 and s_2 were ranked by their preferred New York jobs higher than students matched to those jobs, an instability would now exist, since the couple would prefer to take the two New York jobs, and the New York hospitals would prefer to have s_1 and s_2.

Notice that to describe this kind of instability, we are implicitly proposing a modification of the basic model of agents in the market. A couple consists of a pair of students who have a single preference ordering over *pairs* of positions. Part of the problem with the couples algorithm just described is that it did not permit couples to state their preferences over pairs of positions. Starting in the mid-1980s, modifications were made so that couples could for the first time express such preferences within the framework of the centralized matching scheme. However the following theorem shows that the problem goes deeper than this.

Theorem 5.11 (Roth, Sotomayor). *In the hospital–intern problem with couples, the set of stable matchings may be empty.*

Proof: Consider the market with hospitals $H = \{H_1, H_2, H_3, H_4\}$ each of which offers exactly one position and each of which has strict preferences

Table 5.1. *Preferences of hospitals and couples*

Hospitals' rank orders				Couples' rank orders	
H_1	H_2	H_3	H_4	$\{s_1, s_2\}$	$\{s_3, s_4\}$
s_4	s_4	s_2	s_2	$H_1 H_2$	$H_4 H_2$
s_2	s_3	s_3	s_4	$H_4 H_1$	$H_4 H_3$
s_1	s_2	s_1	s_1	$H_4 H_3$	$H_4 H_1$
s_3	s_1	s_4	s_3	$H_4 H_2$	$H_3 H_1$
				$H_1 H_4$	$H_3 H_2$
				$H_1 H_3$	$H_3 H_4$
				$H_3 H_4$	$H_2 H_4$
				$H_3 H_1$	$H_2 H_1$
				$H_3 H_2$	$H_2 H_3$
				$H_2 H_3$	$H_1 H_2$
				$H_2 H_4$	$H_1 H_4$
				$H_2 H_1$	$H_1 H_3$

over students $S = \{s_1, s_2, s_3, s_4\}$ as given in Table 5.1. The students consist of two married couples, $\{s_1, s_2\}$ and $\{s_3, s_4\}$. Each couple has strict preferences over ordered pairs of hospitals, as given in Table 5.1. Thus couple $\{s_1, s_2\}$ has as its first choice that s_1 be matched with H_1 and s_2 with H_2, and has as its last choice that s_1 be matched with H_2 and s_2 with H_1. The 24 individually rational matchings of students to hospitals are listed in Table 5.2, along with the reason that each such matching is unstable. Thus matching 1, which assigns student s_i to hospital H_i, $i = 1, ..., 4$, is unstable because both hospital H_2 and couple $\{s_3, s_4\}$ would prefer that student s_4 be matched with H_2. (This follows since H_2 prefers s_4 to s_2, and $\{s_3, s_4\}$ prefers $H_3 H_2$ to $H_3 H_4$.)

Note that the emptiness of the set of stable outcomes here is not a "knife-edge" phenomenon. The example would be completely unchanged, for instance, if any preferences for H_2 that kept student s_4 as the first choice were substituted. It is an open question whether there exist plausible restrictions on the preferences of the couples that would insure that stable matchings always exist.

The observation that the couples algorithm used prior to the mid-1980s was a particularly instability-prone mechanism, and that when couples are present in the market some instabilities may be unavoidable, goes a long way toward explaining the lack of participation of married couples in the centralized market, in contrast to the high rates of voluntary par-

Table 5.2. *Every matching is unstable*

Matching	H_1	H_2	H_3	H_4	Unstable with respect to
1	s_1	s_2	s_3	s_4	s_4, H_2
2	s_1	s_2	s_4	s_3	s_4, H_2
3	s_1	s_3	s_2	s_4	s_2, H_4
4	s_1	s_3	s_4	s_2	s_4, H_1
5	s_1	s_4	s_2	s_3	s_2, H_4
6	s_1	s_4	s_3	s_2	s_4, H_1
7	s_2	s_1	s_3	s_4	s_4, H_1
8	s_2	s_1	s_4	s_3	s_4, H_2
9	s_2	s_3	s_1	s_4	s_2, H_4
10	s_2	s_3	s_4	s_1	s_4, H_1
11	s_2	s_4	s_1	s_3	s_2, H_4
12	s_2	s_4	s_3	s_1	s_4, H_1
13	s_3	s_1	s_2	s_4	s_4, H_2
14	s_3	s_1	s_4	s_2	s_2, H_3
15	s_3	s_2	s_1	s_4	s_2, H_4
16	s_3	s_2	s_4	s_1	s_2, H_3
17	s_3	s_4	s_1	s_2	s_1, H_1
18	s_3	s_4	s_2	s_1	s_2, H_1
19	s_4	s_1	s_2	s_3	s_4, H_2
20	s_4	s_1	s_3	s_2	s_2, H_3
21	s_4	s_2	s_1	s_3	s_2, H_4
22	s_4	s_2	s_3	s_1	s_2, H_3
23	s_4	s_3	s_1	s_2	s_3, H_3
24	s_4	s_3	s_2	s_1	s_4, H_4

ticipation of single medical school graduates. This does not appear to be an easy problem to fix, particularly as subsequent work (see Section 5.8) has shown that for large markets containing married couples the problem of even determining if a stable matching exists may be computationally infeasible.

5.4.4 *The geographic distribution of interns*

We remarked in Section 1.1 that many rural hospitals have difficulty filling their posts, and that the posts they do fill are largely staffed by graduates of foreign medical schools. To what extent this may adversely affect the actual or perceived quality of medical care in rural areas of the country has been a source of concern in the medical community. The question has been raised whether this problem could be addressed by altering the NIMP procedure so as to alter the distribution of interns to hospitals.

It is easy to see why this looks like an attractive method to address the issue. After all, the vast majority of interns (who are not married to other graduating medical students) go where they are told to go by the matching procedure. So if the algorithm could be altered to suggest a more desirable distribution of interns to hospitals, maybe the distributional problem could be ameliorated.

However in view of our examination of the history of this market to date, we can see there are constraints on what can be done in this way. High voluntary rates of orderly participation are closely associated with the stability of the matching mechanism. Any unstable matching that might be proposed gives some students and hospitals the incentive to find mutually preferable matches, and the history of the market shows they are prepared to act on these incentives. So any redistribution of interns to hospitals that is to be accomplished in a voluntary manner must be done without sacrificing the stability of the suggested matching. The following two theorems show that for the particular problem at hand, this constraint does not leave any room to maneuver.

Theorem 5.12. *When all preferences over individuals are strict, the set of students employed and positions filled is the same at every stable matching.*

The proof is immediate via the similar result for the marriage problem (Theorem 2.22) and Lemma 5.6.

So any hospital that fails to fill all of its positions at some stable matching will not be able to fill any more positions at any other stable matching. The next result shows that not only will such a hospital fill the same number of positions, but it will fill them with exactly the same interns at any other stable matching.

Theorem 5.13 (Roth). *When preferences over individuals are strict, any hospital that does not fill its quota at some stable matching is assigned precisely the same set of students at every stable matching.*

We will defer the proof of this theorem until after the proof of Lemma 5.25, since that will allow a simpler treatment.

In view of Theorems 5.12 and 5.13, it seems most unlikely that the shortage of interns at rural hospitals, and particularly of interns educated at American medical schools, can be substantially changed simply by alterations in the matching algorithm. Indeed, any voluntary system of organizing the market (so long as it results in stable matchings) will produce the same result, so long as the relative numbers of available positions and eligible students remain the same, unless there is some systematic change in the preferences of students for rural hospitals. (Attempts to produce such changes in preferences are occasionally considered, for example by a program of forgiving loans of students who accept rural positions.)

5.4.5 *Strategic questions*

We turn now to the kind of strategic question that caused the 1950 "trial run" algorithm to be abandoned in favor of the NIMP algorithm: Is it always in the agents' best interest to state their true preferences? From the impossibility theorem for the special case of the marriage market (Theorem 4.4) we know that no stable matching mechanism can have this property for all agents. But in the marriage market we observed that a mechanism that produced the optimal stable matching for one side of the market made it a dominant strategy for agents on that side to state their true preferences (Theorem 4.7). We might therefore hope that the parallel result holds for the hospital intern market, in which case the NIMP algorithm, which produces the hospital-optimal stable matching, would make it a dominant strategy for hospitals to state their true preferences. However this is not the case. As the next theorem shows, Theorem 4.7 is one of those results that does not generalize from the case of one-to-one matching to the case of many-to-one matching.

Theorem 5.14 (Roth). *No stable matching mechanism exists that makes it a dominant strategy for all hospitals to state their true preferences.*

The proof makes use of the same example used to prove Theorem 5.10. (Since a stable matching mechanism that gives all hospitals a dominant strategy would do so for all examples, it is sufficient to show for one example that no such stable matching mechanism exists.)

Proof: Let the sets H and S of agents, the quotas, and the true preferences P, be those of the example used in the proof of Theorem 5.10. Then it is straightforward to verify that when all agents state their true preferences, the set of stable matchings contains a *unique* matching $\mu = \mu_H = \mu_S$. Thus any stable matching procedure must select the outcome μ, and so hospital H_1 receives the assignment $\mu(H_1) = \{s_3, s_4\}$; that is, it is assigned its third and fourth choice students.

Suppose now that hospital H_1 were to state instead the (false) preference ordering $P'(H_1)$ given by $P'(H_1) = s_1, s_4, H_1$, and that all other agents were to state their true preferences, so that the vector of stated preferences is $P' = (P'(H_1), P(H_2), P(H_3), P(s_1), ..., P(s_4))$. It is straightforward to verify that the set of stable matchings with respect to P' also contains a unique matching, which is the matching μ' given by $\mu'(H_1) = \{s_1, s_4\}$, $\mu'(H_2) = \{s_2\}$, $\mu'(H_3) = \{s_3\}$. Thus any stable matching procedure must select the matching μ'. Since H_1 thus receives the assignment $\mu'(H_1) = \{s_1, s_4\}$, which it prefers to $\mu(H_1) = \{s_3, s_4\}$, it does better by stating $P'(H_1)$ than by stating its true preference $P(H_1)$. This completes the proof.

An immediate corollary of the proof is that Theorems 4.10 and 4.11 also do not generalize from the special case of the marriage model. That is, we have

Corollary 5.15. *In the college admissions model, the conclusions of Theorems 4.10 and 4.11 for the marriage model do not hold. A coalition of agents (in fact, even a single agent) may be able to misrepresent its preferences so that it does better than at any stable matching.*

Although Theorem 5.14 shows that no stable matching mechanism gives hospitals a dominant strategy, the situation of students is as in the marriage problem. That is, we have the following result.

Theorem 5.16 (Roth). *A stable matching procedure that yields the student-optimal stable matching makes it a dominant strategy for all students to state their true preferences.*

Proof: If the theorem were not true, there would be some hospital-intern market in which a student could profitably misrepresent his or her preferences, and this would imply that he or she could also profitably misrepresent preferences in the related marriage market, in contradiction to Theorem 4.7.

So the claim in the medical literature that no hospital or student can ever do better than to state true preferences when participating in the NIMP algorithm is not correct. However, if the procedure were reversed so that it produced the student-optimal rather than the hospital-optimal stable matching, then it would be a dominant strategy for the students (but not for the hospitals) to state their true preferences.

As in the case of the marriage market, these results do little to help us identify "good" strategies for either the students or the hospitals. Neither have dominant strategies under the NIMP procedure, so they face complex decision problems. And we cannot even say as much about equilibria, even under the assumption of complete information, as we could for the marriage market, since there are lots of Nash equilibria, and no easy way to distinguish among them because the lack of dominant strategies prevents us from eliminating unreasonable equilibria as in Section 4.4.2.

However since Theorem 5.16 establishes that the *student*-optimal stable mechanism makes it a dominant strategy for students to state their true preferences, we might hope to have at least a one-sided generalization of Theorem 4.16, which would say that every equilibrium of the student-optimal stable mechanism at which students state their true preferences is stable with respect to the true preferences. But this is another result that fails to generalize, even in this partial way, from the special case of the marriage model. Again, the result is a corollary of the proof of Theorem 5.14.

Corollary 5.17. *In the college admissions model, the conclusions of Theorem 4.16 for the marriage model do not hold, even for the student-optimal stable mechanism. When all students state their true preferences, there may be equilibria of the student-optimal stable mechanism that are not stable with respect to the true preferences.*

Proof: Observe that in the proof of Theorem 5.14, the stated strategies for the hospitals are in equilibrium when the student-optimal stable mechanism is used.

In general, although there are equilibrium misrepresentations that yield stable matchings with respect to the true preferences, there are also equilibrium misrepresentations that yield any individually rational matching, stable or not.

Theorem 5.18 (Roth). *There exist Nash equilibrium misrepresentations under any stable matching mechanism that produce any individually rational matching with respect to the true preferences.*

Proof: Let μ be any individually rational matching (including any stable matching) with respect to the true preferences, and let every student s submit a preference list in which $\mu(s)$ is the only acceptable alternative, and every hospital H submit a list in which only the students in $\mu(H)$ are acceptable. Then μ will be the only stable matching under the stated preferences, so it will be the result of the matching mechanism, and no agent can improve his or her assignment by stating another preference so long as all the other agents submit the preference list described here.

It should be clear why this is not a very illuminating result. On the one hand, the set of individually rational matchings includes all remotely plausible matchings, so the set of matchings that can be obtained by equilibrium misrepresentations is uninformatively large. On the other hand, the equilibrium strategies used in the proof require an enormous amount of coordination among the agents, which in the kinds of economic environments of interest here could not in general be sustained. So the fact that all individually rational matchings can be obtained at equilibrium does not mean that they can all be obtained at some reasonable notion of equilibrium, or without the unrealistic assumption that the agents know one anothers' preferences in detail. It would be nice to obtain a sharper result using a more restrictive notion of equilibrium, as we were able to do for the marriage model in Theorem 4.16 in which we could rule out dominated strategies for the men and get only stable matchings at the resulting equilibria.

This is particularly so since the evidence from observations of the hospital–intern market very strongly suggests that the market is achieving stability (or at least a sufficiently close approximation to eliminate the kind of chaotic recontracting found in the market before the NIMP procedure was introduced), and since this evidence is confirmed by examination of other markets (see Section 5.5). Perhaps further theoretical progress in the direction of more fully explaining this may come from studying models of incomplete information, such as those in Section 4.5, that better take into account the kind of information actually available to students and hospitals.

In the meantime, the strategic problems associated with this market are understood well enough to state confidently only that they are complex, since neither students nor hospitals have any dominant strategy under the rules by which the market is presently organized. The hypothesis of equilibrium behavior in this market is confronted squarely with the problems of coordination and information discussed in Chapter 4. Consequently, many open problems remain.

5.5   Some further remarks on empirical matters

There are several reasons why we have devoted some attention, in a book primarily concerned with mathematical theory, to the way American physicians get their first jobs. One reason is to suggest why we think that the body of theory developed here has empirical content. Another reason is simply to give readers an idea of what empirical work connected with theory of this kind might look like. And a third reason is because it seems likely that the lessons learned from the rather special market for American medical interns may generalize to a much wider variety of entry level labor markets and other matching processes. Although the scope of this book does not permit us to elaborate at much length on any of these points, a few additional comments may not be out of place.

Regarding the empirical content of the theory, we have laid great weight in our explanation of the history of the medical market on the fact that the centralized market mechanism introduced in 1951 is a stable matching mechanism, and on the fact that the growing numbers of married couples in the market introduce instabilities. It might be objected that these are coincidental features of the market, and that the true explanations of the rates of participation, for example, lie elsewhere.

For example, it might be postulated that *any* centralized market organization would have solved the problems experienced prior to 1951, and that the difficulties with having married couples in the market have less to do with instabilities of the kind dealt with here than with the difficulties that young couples have in making decisions.

Ideally, we would like to be able to conduct carefully controlled experiments designed to distinguish between any such alternate hypotheses. (And laboratory experimentation is indeed becoming more common in economics. See, e.g., Roth (1987, 1988a) for some general discussions of experimentation in economics, and Harrison and McCabe (1989) for a preliminary experimental study of the operation of the deferred acceptance algorithm in the marriage problem.) But for theories involving the histories of complex natural organizations, we often have to settle for finding "natural experiments" that let us distinguish as well as we can between competing hypotheses.

A very nice natural experiment involving these matters can be found when we look across the Atlantic Ocean and examine how new physicians in the United Kingdom obtain their first jobs. The following brief description is taken from Roth (1989b).

5.5.1 *Regional markets for physicians and surgeons in the United Kingdom*

Around the middle of the 1960s, the entry level market for physicians and surgeons in England, Scotland, and Wales began to suffer from some of the same acute problems that had arisen in the American market in the 1940s and 1950s. Chief among these was that the date of appointment for "preregistration" positions (comparable to American internships, and required of new medical school graduates) had crept back in many cases to years before the date a student would graduate from medical school. The market for these positions is regional rather than national, and this problem occurred more or less in the same way in many of the regional markets. (These regional markets have roughly 200 positions each, compared to roughly 20,000 in the American market in recent years.)

The British medical authorities were aware of the experience of the American market, and in many of the regional markets it was decided to introduce a centralized market mechanism using a computerized algorithm to process preference lists obtained from students and hospitals, modeled loosely after the American system, but adapted to local conditions. Most of these algorithms were not stable matching mechanisms, and it appears that a substantial majority of those that were not failed to solve the problems they were designed to address, and were eventually abandoned. (Before being abandoned, at least some experienced serious incentive problems, the evidence being a lack of voluntary participation, or a variety of unstraightforward strategic behavior.) As far as can so far be determined, only two algorithms based on stable matching mechanisms were introduced (in Edinburgh and Cardiff). These were both largely successful, and remain in use to this day. The similarity of the British experience in markets with unstable mechanisms to the American situation prior to 1951, and the similarity of the British experience in the markets based on stable mechanisms to the American experience after 1951, support the argument that stability plays at least something like the role we have attributed to it.

Since most of our discussion has so far concerned stable mechanisms, it may be illuminating to consider in more detail the experience with unstable mechanisms in some of the U.K. markets. We turn to this next.

5.5.1.1 Some unstable mechanisms: the experience with priority matching

This section considers two closely related matching schemes developed in Newcastle and Birmingham in the late 1960s and subsequently abandoned. In these schemes, a student's ranking of a particular hospital program was combined with the consultant (supervising) physician's ranking of that student to produce a "priority" for that student to be employed by that consultant. The overall matching of students with consultants was determined by making the individual matches of students with positions in order of priority. That is, the first step of the algorithm was to make all first priority matches. Then consultants with unfilled positions and students still needing jobs were scanned to identify second priority matches, and so on.

The Newcastle and Birmingham schemes each used the product of the student's ranking of the consultant and the consultant's ranking of the student as the basis for the priorities. If a consultant and student each ranked one another first – a "(1, 1) match" – they had a priority of one. If the consultant ranked the student first but the student ranked the consultant second – a (1, 2) match – they had a priority of two, as did a consultant who ranked a student second but was ranked first by the student – a

(2, 1) match. The two schemes differed in how they broke ties. In Birmingham ties were broken in the consultant's favor, so a (1, 2) match would have a higher priority than a (2, 1) match. In Newcastle ties were broken in the student's favor. The following example illustrates the algorithms and also proves the following proposition.

Proposition 5.19.   *Each of these schemes may produce unstable matchings.*

Example 5.20
The rank orderings of the consultants and students are as follows. Each consultant has only one position.

$$P(C_1) = s_1, \ldots \qquad\qquad P(s_1) = C_1, \ldots$$
$$P(C_2) = s_1, s_3, s_2, s_4, s_5, s_6 \qquad P(s_2) = C_2, C_1, C_3, C_4, C_5, C_6$$
$$P(C_3) = s_3, s_4, \ldots \qquad\qquad P(s_3) = C_4, C_3, \ldots$$
$$P(C_4) = s_4, s_3, \ldots \qquad\qquad P(s_4) = C_3, C_4, \ldots$$
$$P(C_5) = s_1, s_2, s_5, s_3, s_4, s_6 \qquad P(s_5) = C_1, C_2, C_5, C_3, C_4, C_6$$
$$P(C_6) = s_2, s_5, \ldots \qquad\qquad P(s_6) = C_5, C_2, \ldots$$

The Birmingham algorithm makes the following matches (the priority is indicated in parentheses after each set of matches): $C_1 s_1$ (1, 1), $C_3 s_3$ and $C_4 s_4$ (1, 2), $C_2 s_2$ (3, 1), $C_5 s_6$ (6, 1), $C_6 s_5$ (2, 6). This outcome is unstable because C_5 and s_5 are one another's third choices, but in the Birmingham match they are not matched to each other, but are each matched to their *sixth* choice. The Newcastle algorithm makes the following matches: $C_1 s_1$ (1, 1), $C_3 s_4$ and $C_4 s_3$ (2, 1), $C_2 s_2$ (3, 1), $C_5 s_6$ (6, 1), $C_6 s_5$ (2, 6). This outcome is also unstable with respect to C_5 and s_5.

So far the example has been analyzed as if the agents all state their true preferences. Before considering the incentives agents may have to do otherwise, it will be illuminating to examine the history of these matching systems after their introduction, and how they failed and were abandoned. The following description of the events leading to the abandonment of the Newcastle scheme is from a letter by Dr. John Anderson, the postgraduate dean there.

Every six months there is a small number of posts that are left unfilled. This is the background to our problems, and this imbalance between local graduates and posts explains why the computerized scheme failed. Understandably, consultants in the periphery of the region were anxious to fill their posts as quickly as possible and often entered into private arrangements with undergraduates.... The practice of making private arrangements outwith the computer match scheme gradually spread to the Teaching Hospitals. Those who stuck rigidly to the scheme often found that they were left without any housemen to appoint, as there was no way of preventing these private arrangements and no sanctions could be introduced against those who operated outside the scheme.

In the late 70s and early 80s an increasing number of problems cropped up, mainly concerning conflicts between private arrangements and the formal application procedure. There was a feeling that the computer scheme was an impersonal mechanism which inhibited personal contact between students and consultants and *shortly before the scheme was discarded we found that in up to 80% of cases students and consultants only used the computer to indicate a first preference.... The main reason for the abandonment of the scheme, therejore, was that there were problems in getting students and consultants to participate in an orderly way, and this led to those who rigidly observed the requirements of the scheme to be penalised.* (emphasis added)

The experience in Birmingham was similar. There the centralized procedure, which was initiated in 1966 for a limited group of hospitals, failed after a few years, was resumed on a larger scale in 1971, failed once more, was restarted again around 1978, and was finally abandoned again around 1981. A recurrent theme in accounts of the failures of these schemes is that after the centralized matching had been in use for a short while, increasing numbers of jobs began once again to be arranged privately in advance between consultants and students, and that this worked to the detriment of those who tried to participate in the scheme without prior arrangement. To understand this phenomenon, consider now the incentives that these priority procedures give to the agents. For this purpose, consider again Example 5.20. To make the example clear, suppose consultants C_1 through C_4 are in the most desirable teaching hospital, C_5 is in the next most desirable regional hospital, and C_6 is in a relatively undesirable rural hospital. Similarly, suppose students s_1 through s_5 are all top graduates of their medical school, while s_6 has a less distinguished record. Then C_5 is disappointed to learn that his new junior house officer will be s_6, all the more when he learns that student s_5, whom he liked reasonably well, is quite unhappy with his own appointment, and would have preferred to work for C_5. If C_5 had submitted a rank ordering on which s_5 was his first choice, they would have been matched, as would also have been the case if s_5 had submitted a rank ordering on which C_5 was his first choice. So the example shows there may be incentives for both students and consultants to submit rank orderings different from their true preferences.

Furthermore, these unstable priority ranking systems allow such incentives to build upon one another, so that as more agents adapt their submitted rank orderings to improve their matches, the greater is the incentive for other agents to do so. To see this, suppose C_5 in Example 5.20 resolves not to suffer the same fate the following year. He therefore approaches one of the good students in the next year's class, in advance of the formal match, and suggests that they mutually agree to be matched, which they will accomplish by ranking one another first in the formal match. The student, chastened by the experience of s_5 the previous year, is receptive. (In the specific context of these relatively small markets, both parties to such an agreement can be confident that it will be carried out.) Now consider the situation in the formal match, when a number of positions have been prearranged to be (1, 1) matches. Suppose students t_1, t_2, and t_3 have made such arrangements with consultants C_3, C_4, and C_5, but consultant C_2, not knowing this, submits his true rank ordering, which is t_1, t_2, t_3, t_4, t_5, ..., and t_4 submits his true rank ordering C_3, C_4, C_5, C_2, ... Although C_2 doesn't know it, t_4 is his highest ranking student who is actually available, and C_2 is t_4's most preferred available consultant. But the product of their rankings is 16, and C_2 could well end up with his *fifteenth* choice student. So when some matches have been prearranged, those not in the know, students as well as consultants, stand to do very poorly if they do not also prearrange their matches. Furthermore, when an agent's top n choices have all arranged to indicate only a first preference in the formal match (as in the preceding quote from Anderson), then the agent can do no better in the match than to personally reach such an agreement with his or her $n+1$st choice. So we have proved the following proposition.

Proposition 5.21 (Roth). *It is not a dominant strategy for any agent to submit his or her true preferences in these priority matching systems. Furthermore, there are multiple equilibria at which all agents submit only a first choice.*

In fact, the proposition does not capture the full strength of what has been proved. Under priority matching, a student and consultant who rank one another first will be matched regardless of what the other agents do. So the problems of coordination that afflict many equilibria do not arise here. Pairs of agents may secure their part of the equilibrium by private arrangement. These results thus go a long way toward explaining both why a high percentage of appointments were soon arranged in advance under these unstable mechanisms, and why this worked to the disadvantage of those who tried to arrange employment through these priority-based formal match procedures.

5.5.1.2 Some modeling issues
The nature of this kind of empirical investigation is of course very different from the purely mathematical investigation of abstract cases. Particular models adapted to the institutional details of the markets in

question must be considered (just as considering instabilities involving married couples required us to extend the basic hospital–intern model). To give a bit of the flavor of this, one example comes to mind.

One of the stable matching procedures was introduced in a region of Scotland, where in keeping with previous custom, certain kinds of hospital programs were permitted to specify that they did not wish to employ more than one female physician at any time. A program taking advantage of this option might submit a preference list on which several women graduates were highly ranked, but nevertheless stipulate that no more than one of these should be assigned to it. In analyzing such a model, it is of course necessary to consider whether the introduction of such "discriminatory quotas" influences the existence of stable matchings. We leave as an exercise for the reader to construct the variation on the basic model needed to address this question, and to prove the following proposition.

Proposition 5.22 (Roth).   *In the hospital–intern model with discriminatory quotas, the set of pairwise stable matchings is always nonempty.*

Another modeling issue that we have not mentioned but that proved to be important in the study of the U.K. markets in contrast to the American market is that in the U.K., each student must seek *two* positions, one in medicine and one in surgery. So the formal models used to study such markets must be many-to-two, instead of many-to-one as in the college admissions model. It is not too difficult to show that when all preferences are responsive, the set of stable matchings is still nonempty in such models (see Roth 1989b), but the following proposition shows that the relationship between (pairwise) stability and group stability established for many-to-one matching in Lemma 5.5 need no longer hold.

Proposition 5.23 (Roth).   *In many-to-many matching models with responsive preferences, a stable matching need not be group stable, and need not even be Pareto optimal.*

The proof is via the following example.

Example 5.24: Many-to-many matching with responsive preferences (Roth)
There are four firms and four workers, each with a quota of two, with (responsive) preferences as follows.

$$P^{\#}(w_1) = \{F_1, F_2\}, \{F_1, F_3\}, \{F_1, F_4\}, \{F_2, F_3\}, \{F_2, F_4\}, \{F_3, F_4\},$$
$$\{F_1\}, \{F_2\}, \{F_3\}, \{F_4\}$$

$$P^{\#}(w_2) = \{F_2, F_1\}, \{F_2, F_4\}, \{F_2, F_3\}, \{F_1, F_4\}, \{F_1, F_3\}, \{F_4, F_3\},$$
$$\{F_2\}, \{F_1\}, \{F_4\}, \{F_3\}$$

$$P^{\#}(w_3) = \{F_3, F_4\}, \{F_3, F_1\}, \{F_3, F_2\}, \{F_4, F_1\}, \{F_4, F_2\}, \{F_1, F_2\},$$
$$\{F_3\}, \{F_4\}, \{F_1\}, \{F_2\}$$

$$P^{\#}(w_4) = \{F_4, F_3\}, \{F_4, F_2\}, \{F_4, F_1\}, \{F_3, F_2\}, \{F_3, F_1\}, \{F_2, F_1\},$$
$$\{F_4\}, \{F_3\}, \{F_2\}, \{F_1\}$$

$$P^{\#}(F_1) = \{w_4, w_3\}, \{w_4, w_2\}, \{w_4, w_1\}, \{w_3, w_2\}, \{w_3, w_1\},$$
$$\{w_2, w_1\}, \{w_4\}, \{w_3\}, \{w_2\}, \{w_1\}$$

$$P^{\#}(F_2) = \{w_3, w_4\}, \{w_3, w_1\}, \{w_3, w_2\}, \{w_4, w_1\}, \{w_4, w_2\},$$
$$\{w_1, w_2\}, \{w_3\}, \{w_4\}, \{w_1\}, \{w_2\}$$

$$P^{\#}(F_3) = \{w_2, w_1\}, \{w_2, w_4\}, \{w_2, w_3\}, \{w_1, w_4\}, \{w_1, w_3\},$$
$$\{w_4, w_3\}, \{w_2\}, \{w_1\}, \{w_4\}, \{w_3\}$$

$$P^{\#}(F_4) = \{w_1, w_2\}, \{w_1, w_3\}, \{w_1, w_4\}, \{w_2, w_3\}, \{w_2, w_4\},$$
$$\{w_3, w_4\}, \{w_1\}, \{w_2\}, \{w_3\}, \{w_4\}$$

Each agent's preferences P over individuals can be read from the last four (singleton) entries in his or her preference list $P^{\#}$ over sets. The matching μ that matches each agent to his or her fourth choice set of agents (emphasized in boldface in the preferences – i.e., $\mu(w_1) = \{F_2, F_3\}$, $\mu(F_1) = \{w_3, w_2\}$, etc.) is (pairwise) stable. To see this, note that for each w_i, all improvements on $\mu(w_i)$ involve firm F_i, but no firm F_i is interested in dealing with w_i alone, since w_i is firm F_i's last choice individual. But μ isn't in the core, and isn't Pareto optimal, since it is dominated by the matching μ' that gives each agent his or her *third* choice set of agents. (In each case, the responsive preferences have been chosen so that an agent prefers to be matched with the set consisting of his or her first and fourth choice individuals rather than to his or her second and third choices.)

It is easy to see from this example why the arguments of Lemma 5.5 do not carry over to the case of many-to-many matching. In the case of many-to-one matching, if a matching μ was group unstable, it was possible to find a worker–firm (student–college) pair who blocked it. However in the present example, μ' matches every worker with a firm he or she likes more *and* with a firm he or she likes less than his or her mates at μ, and similarly for firms. So the only effective coalitions are larger than worker–firm pairs.

Example 5.24 shows that in many-to-many matching, even when no pair of agents can together arrange to do better than a given matching, there might be a larger coalition who by rearranging job assignments among many agents could obtain preferred assignments for all its members. Needless to say, identifying and organizing large coalitions may be more difficult than making private arrangements between two parties, and the experience of those regional markets in the United Kingdom that are built around stable mechanisms suggests that pairwise stability is still of primary importance in these markets.

Regarding directions for future empirical work, we remark that the two studies discussed here (Roth 1984a, 1989b) are both part of a line of work that seeks to identify markets in which it is possible to establish a particularly close connection between the observed market outcome and the set of stable outcomes. This connection can be made so closely because the markets in question used computerized matching procedures that can be examined to determine the precise relationship between the submitted preferences and the market outcome. But the kind of theory developed here is by no means limited to such markets, and as more becomes known about the behavior of other entry level labor markets, for example, we should be better able to associate certain phenomena with markets that achieve stable outcomes, and other phenomena with markets that achieve unstable outcomes. In this way it should be possible to extend the empirical investigation of the predictions of this kind of theory to two-sided matching markets that are operated in a completely decentralized manner.

An interesting intermediate case that has been described in Mongell (1988) and Mongell and Roth (1989) concerns the procedures by which the social organizations known as sororities, which operate on many American campuses, are matched each year with new members. A centralized procedure is employed that in general would not lead to a stable matching, but because the agents in that market respond to the incentives that the procedure gives them not to state their full true preferences, much of the actual matching in that market is done in a decentralized aftermarket. In the data examined by Mongell and Roth, the strategic behavior of the agents led to stable matches.

Finally, it should be able to go without saying that empirical observations don't "prove" a mathematical theory any more than a mathematical theory by itself can "prove" an assertion about market behavior. Nevertheless, at least one of the authors would feel very differently about the theory presented here if the weight of the empirical evidence were different.

We turn now from empirical considerations to developing the theoretical properties of the basic college admissions model.

5.6 Comparisons of stable matchings in the college admissions model

We have seen that the college admissions model of many-to-one matching is different from the marriage model of one-to-one matching in a number of important respects. So far, the most important differences have made themselves felt in terms of regularities of the marriage model that did not carry over to the college admissions model; for example, in the marriage model the optimal stable matching for one side is Pareto optimal for that side, and it is a dominant strategy for agents on one side to state their true preferences when a stable mechanism that produces the optimal stable matching for their side is used, but Theorems 5.10 and 5.14 show that neither of these results holds for the colleges in the college admissions model.

In this section we turn to the opposite kind of difference between the two models, and examine regularities that appear in the more general college admissions model that have no parallel in the special case of the marriage model. We continue to assume throughout that students and colleges have complete and transitive preferences over individuals, and that colleges have responsive preferences over groups of students. Throughout this section we will further assume, unless stated otherwise, that all preferences over *individuals* are strict. We explicitly do *not* assume that colleges preferences over groups are strict: Colleges may be indifferent between distinct groups of students.

5.6.1 *The main results: an example*

To introduce the main results by way of an example, suppose one of the colleges, say college C, gives an entrance exam and evaluates students according to their scores on this exam, and evaluates entering classes according to their *average* score on the exam. (So even when we assume no two students have exactly the same score on the exam, so that college C's preferences over individuals are strict, it does not have strict preferences over entering classes, since it is indifferent between two entering classes with the same average score.) Then different stable matchings may give college C different entering classes. However for this example our results imply that *no two distinct entering classes that college C could have at stable matchings will have the same average exam score.* Furthermore, for any two distinct entering classes that college C could be assigned at stable matchings, we can make the following strong comparison. Aside from the students who are in both entering classes, *every student in one* *of the entering classes will have a higher exam score than any student in the other entering class.*

More generally, the results say that in a college admissions problem in which all preferences over individuals are strict, no college will be indifferent between any two (different) groups of students that it enrolls at stable matchings. Furthermore, for every pair of stable matchings, each college will prefer every student who is assigned to it at one of the two matchings to every student who is assigned to it in the second matching but not the first. These results are mathematically unusual, and also have significant implications for the use of these models in understanding empirical economic phenomena.

The manner in which they are mathematically unusual can be understood by the following observation. The first result, for example, can be rephrased to say that if a given matching is stable (and hence in the core; see Section 5.7), and if some college is indifferent between the entering class it is assigned at that matching and a different entering class that it is assigned at a different matching, then the second matching is *not* in the core. We thus have a way of concluding that an outcome is not in the core, based on the direct examination of the preferences of only *one* agent (the college). Since the definition of the core involves preferences of coalitions of agents, this is quite surprising, and in fact we know of no comparable results concerning the core of a game (except in the trivial case of classes of games that have at most a single outcome in the core).

The significance of these results for using this kind of model to study observable markets and matching processes has to do with the modeling questions that arise whenever an organization is modeled as an individual agent rather than a collection of agents. Although it may often be an acceptable approximation to model individual actors within (an appropriately small part of) an organization as sharing the same preferences over candidates (e.g., when the information about candidates consists primarily of things like standardized exam scores and grades), more is involved in assuming they have the same preferences over entering classes. For example, in the study of the American hospital–intern labor market, the agents on the institution side of the market were modeled as being the programs within a hospital that offer a particular kind of internship. Several physicians are typically associated with each such program, and one can imagine that some might evaluate an entering class by paying most attention to the quality of its best members, whereas others might evaluate entering classes by weighing most heavily the quality of the weakest candidates admitted. This kind of systematic divergence of interests might make it necessary to model in more detail the governance structure of the hospital program, and make it inappropriate to model the program as a single agent. However the results of this section (specifically Corollary 5.28) demonstrate that this kind of divergence of preferences cannot arise in connection with comparison of stable matchings, since for entering classes that could be admitted at stable matchings, the rank ordering of entering classes in terms of their best member, for example, is the same as the rank ordering in terms of their worst member (modulo indifference, as when two potential entering classes have the same individual as the best member, but different worst members). These results may therefore help to explain the success that this kind of model has had in explaining empirical observations.

5.6.2 The main results

Lemma 5.25 (Roth and Sotomayor). *Suppose colleges and students have strict individual preferences, and let μ and μ' be stable matchings for (S, C, P), such that $\mu(C) \neq \mu'(C)$ for some C. Let $\bar{\mu}$ and $\bar{\mu}'$ be the stable matchings corresponding to μ and μ' in the related marriage market. If $\bar{\mu}(c_i) >_C \bar{\mu}'(c_i)$ for some position c_i of C then $\bar{\mu}(c_j) \geq_C \bar{\mu}'(c_j)$ for all positions c_j of C.*

Proof: It is enough to show that $\bar{\mu}(c_j) >_C \bar{\mu}'(c_j)$ for all $j > i$. So suppose this is false. Then there exists an index j such that $\bar{\mu}(c_j) >_C \bar{\mu}'(c_j)$, but $\bar{\mu}'(c_{j+1}) \geq_C \bar{\mu}(c_{j+1})$. Theorem 5.12 implies $\bar{\mu}'(c_j) \in S$. Let $s' \equiv \bar{\mu}'(c_j)$. By the decomposition lemma $c_j \equiv \bar{\mu}'(s') >_{s'} \bar{\mu}(s')$. Furthermore, $\bar{\mu}(s') \neq c_{j+1}$, since $s' >_C \bar{\mu}'(c_{j+1}) \geq_C \bar{\mu}(c_{j+1})$ (where the first of these preferences follows from the fact that for any stable matching $\bar{\mu}'$ in the related marriage market, $\bar{\mu}'(c_j) >_C \bar{\mu}'(c_{j+1})$ for all j). Therefore $c_{j+1} >_{s'} \bar{\mu}(s')$, since c_{j+1} comes right after c_j in the preferences of s' (or any s) in the related marriage problem. So $\bar{\mu}$ is blocked via s' and c_{j+1}, contradicting (via Lemma 5.6) the stability of μ.

We can now give the short proof of Theorem 5.13 that we deferred until the necessary tools were developed.

Proof of Theorem 5.13: Recall that if a college C has any unfilled positions, these will be the highest numbered c_j at any stable matching of the corresponding marriage problem. By Theorem 5.12 these positions will be unfilled at any stable matching, that is, $\bar{\mu}(c_j) = \bar{\mu}'(c_j)$ for all such j, and hence for all j, since the proof of Lemma 5.25 shows that if $\bar{\mu}(c_i) >_C \bar{\mu}'(c_i)$ for some position c_i of C, then $\bar{\mu}(c_j) >_C \bar{\mu}'(c_j)$ for all $j > i$.

Since colleges' preferences over groups are responsive to their preferences over individuals, the following result follows from Lemma 5.25.

Theorem 5.26 (Roth and Sotomayor). *If colleges and students have strict preferences over individuals, then colleges have strict preferences over those groups of students that they may be assigned at stable matchings. That is, if μ and μ' are stable matchings, then a college C is indifferent between $\mu(C)$ and $\mu'(C)$ only if $\mu(C) = \mu'(C)$.*

Proof: If $\mu(C) \neq \mu'(C)$, then (without loss of generality) $\bar{\mu}(c_i) >_C \bar{\mu}'(c_i)$ for some position c_i of C, where $\bar{\mu}$ and $\bar{\mu}'$ are the matchings in the related marriage market corresponding to μ and μ'. By Lemma 5.25, $\bar{\mu}(c_j) \geq_C \bar{\mu}'(c_j)$ for all positions c_j of C. So $\mu(C) >_C \mu'(C)$, by repeated application of the fact that C's preferences are responsive and transitive. [First compare $\mu'(C)$ with a matching that agrees with $\mu(C)$ on position c_1 (in the related marriage problem) and with $\mu'(C)$ for all other positions, then compare this new matching with a matching that agrees with $\mu(C)$ on positions c_1 and c_2, and with $\mu'(C)$ on all other positions, and so on. Responsiveness of preferences determines each pairwise comparison in the resulting chain, and transitivity then assures the desired result.]

Of course the result of Lemma 5.25 and the proof of Theorem 5.26 can be stated without reference to the related marriage problem. Suppose μ and μ' are stable matchings for (S, C, P), and C is a college with $q_C = k$ such that $\mu(C) \neq \mu'(C)$ and $\mu(C) = \{s_1, \ldots, s_k\}$ and $\mu'(C) = \{s_1', \ldots, s_k'\}$, where the students are listed in order of preference, that is, $s_i >_C s_{i+1}$ and $s_i' >_C s_{i+1}'$ for all i. If i is any index such that $s_i >_C s_i'$ then $s_j \geq_C s_j'$ for all $j \in \{1, \ldots, k\}$ (and $s_j >_C s_j'$ for all $j > i$), and $\mu(C) >_C \mu'(C)$ (where the last comparison follows from the proof of Theorem 5.26).

Before proceeding with further results, let us pause to consider what we have learned. Consider a college C with $q_C = 2$ and preferences $P(C) = s_1, s_2, s_3, s_4$. Consider two matchings μ and ν such that $\mu(C) = \{s_1, s_4\}$ and $\nu(C) = \{s_2, s_3\}$. Then without knowing anything about the preferences of students and other colleges, we can conclude that μ and ν cannot both be stable. Note that this is so even though C may either prefer one of $\mu(C)$ or $\nu(C)$ to the other, or be indifferent between them.

Recall that for fixed preferences over individuals, the set of stable (or group stable) matchings is not sensitive to changes in colleges' preferences for groups of students (so long as those preferences remain responsive to the colleges' preferences over individuals). Theorem 5.26 therefore tells us something not only about the preferences each college actually has for groups of students that it may be assigned at stable matchings, but also about all the different preferences for groups that it *could* have, given its preferences over individuals. That is, let $P(C)$ be college C's preferences over individual students. Then Theorem 5.26 tells us that if μ and μ' are both stable matchings, then no preferences that are responsive to $P(C)$ can be indifferent between the groups of students $\mu(C)$ and $\mu'(C)$. The following theorem makes this clearer.

Theorem 5.27 (Roth and Sotomayor). *Let preferences over individuals be strict, and let μ and μ' be stable matchings for (S, C, P). If $\mu(C) >_C \mu'(C)$ for some college C, then $s >_C s'$ for all s in $\mu(C)$ and s' in $\mu'(C) - \mu(C)$. That is, C prefers every student in its entering class at μ to every student who is in its entering class at μ' but not at μ.*

Proof: Consider the related marriage market (S, C', P) and the stable matchings $\bar{\mu}$ and $\bar{\mu}'$ corresponding to μ and μ'. Let $q_C = k$, so that the positions of C are $c_1, \ldots, c_k$. First observe that C fills its quota under μ and μ', since if not, Theorem 5.13 would imply that $\mu(C) = \mu'(C)$. So $\mu'(C) - \mu(C)$ is a nonempty subset of S, since $\mu(C) \neq \mu'(C)$. Let $s' = \bar{\mu}'(c_j)$ for some position c_j such that $s' \notin \mu(C)$. Then $\bar{\mu}(c_j) \neq \bar{\mu}'(c_j)$. By Lemma 5.25 $\bar{\mu}(c_j) >_C \bar{\mu}'(c_j)$. The decomposition lemma implies $c_j >_{s'} \bar{\mu}(s')$. So the construction of the related marriage problem implies $C >_{s'} \mu(s')$, since $\mu(s') \neq C$. Thus $s >_C s'$ for all s in $\mu(C)$ by the stability of μ, which completes the proof.

To illustrate what Theorem 5.27 adds to what we already know, consider again a college C with $q_C = 2$ and preferences $P(C) = s_1, s_2, s_3, s_4$. Consider two matchings μ and ν such that $\mu(C) = \{s_1, s_3\}$ and $\nu(C) = \{s_2, s_4\}$. Then the theorem says that if μ is stable, ν is not, and vice versa.

The following corollary follows immediately from the theorem and the definition of responsive preferences.

Corollary 5.28 (Roth and Sotomayor). *Consider a college C with preferences $P(C)$ over individual students, and let $P^{\#}(C)$ and $P^*(C)$ be preferences over groups of students that are responsive to $P(C)$, (but are otherwise arbitrary). Then for every pair of stable matchings μ and μ', $\mu(C)$ is preferred to $\mu'(C)$ under the preferences $P^{\#}(C)$ if and only if $\mu(C)$ is preferred to $\mu'(C)$ under $P^*(C)$.*

Corollary 5.28 formalizes our introductory comments about why different individuals (having the same preferences over individual students) can be modeled as a single agent, for example college C, in a model of this kind.

The following example illustrates the results of this section.

Let the preferences over individuals be given by

$$P(s_1) = C_5, C_1 \qquad P(C_1) = s_1, s_2, s_3, s_4, s_5, s_6, s_7$$
$$P(s_2) = C_2, C_5, C_1 \qquad P(C_2) = s_5, s_2$$
$$P(s_3) = C_3, C_1 \qquad P(C_3) = s_6, s_7, s_3$$
$$P(s_4) = C_4, C_1 \qquad P(C_4) = s_7, s_4$$
$$P(s_5) = C_1, C_2 \qquad P(C_5) = s_2, s_1$$
$$P(s_6) = C_1, C_3$$
$$P(s_7) = C_1, C_3, C_4$$

and let the quotas be $q_{C_1} = 3$, $q_{C_j} = 1$ for $j = 2, \ldots, 5$. Then the set of stable outcomes is $\{\mu_1, \mu_2, \mu_3, \mu_4\}$ where

$$\mu_1 = \begin{pmatrix} C_1 & C_2 & C_3 & C_4 & C_5 \\ s_1\ s_3\ s_4 & s_5 & s_6 & s_7 & s_2 \end{pmatrix}$$

$$\mu_2 = \begin{pmatrix} C_1 & C_2 & C_3 & C_4 & C_5 \\ s_3\ s_4\ s_5 & s_2 & s_6 & s_7 & s_1 \end{pmatrix}$$

$$\mu_3 = \begin{pmatrix} C_1 & C_2 & C_3 & C_4 & C_5 \\ s_3\ s_5\ s_6 & s_2 & s_7 & s_4 & s_1 \end{pmatrix}$$

$$\mu_4 = \begin{pmatrix} C_1 & C_2 & C_3 & C_4 & C_5 \\ s_5\ s_6\ s_7 & s_2 & s_3 & s_4 & s_1 \end{pmatrix}.$$

Note that these are the only stable matchings, and

$$\mu_1(C_1) \underset{C_1}{>} \mu_2(C_1) \underset{C_1}{>} \mu_3(C_1) \underset{C_1}{>} \mu_4(C_1),$$

for any responsive preferences.

5.6.2.1 When preferences over individuals are not strict

The following example shows that, as in the case of the marriage problem, many of the structural properties of the set of stable matchings depend on strict preferences over individuals. In particular, the conclusions of Theorem 5.27 do not hold in the following example.

Let the preferences over individuals be given by the following preference lists, where two students are enclosed by brackets if the college is indifferent between them. College C_1 has a quota of two, and C_2 has a quota of one.

$$P(C_1) = s_1, [s_2, s_3] \qquad P(s_1) = C_2, C_1$$
$$P(C_2) = [s_2, s_3], s_1 \qquad P(s_2) = C_1, C_2$$
$$ \qquad P(s_3) = C_1, C_2$$

Then

$$\mu = \begin{pmatrix} C_1 & C_2 \\ s_1, s_3 & s_2 \end{pmatrix} \quad \text{and} \quad \mu' = \begin{pmatrix} C_1 & C_2 \\ s_2, s_3 & s_1 \end{pmatrix}$$

are stable and $\mu(C_1) >_{C_1} \mu'(C_1)$, but

$$s_3 \in \mu(C_1), \quad s_2 \in \mu'(C_1) - \mu(C_1), \quad \text{and} \quad s_3 \underset{C_1}{\not>} s_2.$$

5.6.3 *Further results for the college admissions model*

In the marriage problem, individuals with strict preferences over possible mates must trivially have strict preferences over their mates at stable matchings. The results of the previous section show that when all preferences over individuals are strict, colleges also have strict preferences over their entering classes at stable matchings (even though they may be indifferent between other potential entering classes). These results thus establish a further connection between the stable matchings of college admissions and marriage problems.

In this section we will use these results to establish for the college admissions problem several results established for the marriage model in Chapter 2. These results, except for Theorems 5.34 and 5.35, all appear to depend critically on Lemma 5.25. We continue to assume throughout this section that all preferences over individuals are strict. In what follows, we write $\mu >_C \mu'$ to mean $\mu(C) \geq_C \mu'(C)$ for all $C \in C$ and $\mu(C) >_C \mu'(C)$ for some C in C.

Theorem 5.29. *If μ and μ' are stable matchings for (S, C, P) then $\mu >_C \mu'$ if and only if $\mu' >_S \mu$.*

Proof: Suppose that $\mu(C) \geq_C \mu'(C)$ for all C in C and $\mu(C) >_C \mu'(C)$ for some C in C. Using Lemma 5.25 in one direction and the responsiveness of the colleges' preferences in the other direction, we can see that this is equivalent to $\bar{\mu}(c_i) \geq_{c_i} \bar{\mu}'(c_i)$ for all c_i in C' and $\bar{\mu}(c_j) >_{c_j} \bar{\mu}'(c_j)$ for some c_j in C', when $\bar{\mu}$ and $\bar{\mu}'$ are the stable matchings corresponding to μ and μ' for the related marriage market (S, C', P'). This in turn is satisfied if and only if $\bar{\mu} >_{C'} \bar{\mu}'$ and hence, if and only if $\bar{\mu}' >_S \bar{\mu}$ by Theorem 2.13, which implies $\mu' >_S \mu$.

This of course has the following immediate corollary.

Corollary 5.30. *The optimal stable matching on one side of the market (S, C, P) is the worst stable matching for the other side.*

If μ and μ' are matchings we can define

$$\lambda(C) = \mu(C) \quad \text{if } \mu(C) >_C \mu'(C)$$
$$= \mu'(C) \quad \text{otherwise,}$$

$$\lambda(s) = \mu(s) \quad \text{if } \mu'(s) >_s \mu(s)$$
$$= \mu'(s) \quad \text{otherwise.}$$

That is, λ is obtained by assigning to each college C whichever it prefers of $\mu(C)$ and $\mu'(C)$, and assigning to each student s whichever of $\mu(s)$ and $\mu'(s)$ he or she likes less. Note that λ is a matching only if it has the property that $\lambda(s) = C$ if and only if s is contained in $\lambda(C)$, which will not be the case if s is assigned by λ to more than one college. (For arbitrary matchings μ and μ', λ need not be a matching.)

Similarly, define

$$\nu(s) = \mu(s) \quad \text{if } \mu(s) >_s \mu'(s)$$
$$= \mu'(s) \quad \text{otherwise,}$$

$$\nu(C) = \mu(C) \quad \text{if } \mu'(C) >_C \mu(C)$$
$$= \mu'(C) \quad \text{otherwise.}$$

If λ is a matching then, clearly, λ is the least upper bound for $\{\mu, \mu'\}$ under $>_C$ and the greatest lower bound for $\{\mu, \mu'\}$ under $>_S$. As in the marriage market we will denote

$$\lambda = \mu \underset{C}{\vee} \mu' \quad \text{and} \quad \lambda = \mu \underset{S}{\wedge} \mu'.$$

Symmetrically, if ν is a matching,

$$\nu = \mu \underset{S}{\vee} \mu' \quad \text{and} \quad \nu = \mu \underset{C}{\wedge} \mu'.$$

We are going to show that if μ and μ' are stable, then λ and ν are both matchings, and both stable. This fact will imply that the set of stable matchings is a lattice under $>_C$ and under $>_S$.

Theorem 5.31. *Let μ and μ' be stable matchings for (S, C, P). Then λ and ν are stable matchings.*

Proof: Consider the marriage market (S, C', P') related to (S, C, P) and the stable matchings $\bar{\mu}$ and $\bar{\mu}'$ corresponding to μ and μ'. We know that $\bar{\lambda} = \bar{\mu} \vee_{C'} \bar{\mu}'$ is a stable matching for (S, C', P'), by Theorem 2.16. Now observe that if, say, $\lambda(C) = \mu(C)$, then $\bar{\mu}(c) \geq_c \bar{\mu}'(c)$ for all positions c of C by Lemma 5.25, so $\bar{\lambda}(c) = \bar{\mu}(c)$ for all positions c of C. Hence if s is in $\mu(C)$, there is some position c of C such that

$$s = \bar{\lambda}(c). \tag{1}$$

To see that λ is a matching, suppose by the way of contradiction that there are some s in S and C and C' in C with $C \neq C'$ and such that s is contained in both $\lambda(C)$ and $\lambda(C')$. Then, by (1) there exists some position c of C, and some position c' of C', such that $s = \bar{\lambda}(c) = \bar{\lambda}(c')$, which contradicts the fact that $\bar{\lambda}$ is a matching.

The matching λ is stable, for if $s >_C s' \in \lambda(C)$, so, by (1), there is some position c of C such that $s' = \bar{\lambda}(c)$ and $s >_c \bar{\lambda}(c)$. But then $\bar{\lambda}(s) >_s c$, by stability of $\bar{\lambda}$, which implies that $\lambda(s) >_s C$ and (C, s) does not block λ. Similarly we can show that ν is a stable matching.

Translated into algebraic terms, Theorems 5.29 and 5.31 give us the following result.

Corollary 5.32. *The set of stable matchings forms a lattice under the partial orders $>_C$ or $>_S$ with the lattice under the first partial order being the dual to the lattice under the second partial order.*

Finally, we turn to an analogue of the decomposition lemma.

Theorem 5.33. *If μ and μ' are two stable matchings for (S, C, P) and $C = \mu(s)$ or $C = \mu'(s)$, with $C \in C$ and $s \in S$, then if $\mu(C) >_C \mu'(C)$ then $\mu'(s) \geq_s \mu(s)$ (and if $\mu'(s) >_s \mu(s)$ then $\mu(C) \geq_C \mu'(C)$).*

Proof: Consider the marriage market (S, C', P') and the corresponding stable matchings $\bar{\mu}$ and $\bar{\mu}'$. Define $S(\bar{\mu}') = \{s \in S; \; \bar{\mu}'(s) >_s \bar{\mu}(s)\}$, and $C'(\bar{\mu}) = \{c \in C'; \; \bar{\mu}(c) >_c \bar{\mu}'(c)\}$. Similarly define $S(\bar{\mu})$ and $C'(\bar{\mu}')$.

By the decomposition lemma (specifically Corollary 2.21) $\bar{\mu}$ and $\bar{\mu}'$ map $S(\bar{\mu}')$ onto $C'(\bar{\mu})$ and $S(\bar{\mu})$ onto $C'(\bar{\mu}')$. Then, if $\mu(C) >_C \mu'(C)$, Lemma 5.25 implies that $\bar{\mu}(c) \geq_c \bar{\mu}'(c)$ for all positions c of C. Then $c \notin C'(\bar{\mu}')$ for all positions c of C. Then $\bar{\mu}(c)$ and $\bar{\mu}'(c)$ are in $S(\bar{\mu}')$ or $\bar{\mu}(c) = \bar{\mu}'(c)$, for all positions c of C. Since s is matched to some position of C under $\bar{\mu}$ or $\bar{\mu}'$, we have that $\mu'(s) \geq_s \mu(s)$. Now if $s \in S(\bar{\mu}')$, then $\bar{\mu}(s)$ and $\bar{\mu}'(s)$ are in $C'(\bar{\mu})$ by the decomposition lemma. Since s is matched to some position of C under $\bar{\mu}$ or $\bar{\mu}'$, the result follows by Lemma 5.25.

The next two results are the extensions to the college admissions model of Theorems 2.24 and 2.25, concerning the effect of extending agents' preferences, and of adding new agents to the market.

Theorem 5.34 (Gale and Sotomayor). *Suppose $\bar{P} \geq_C P$ and let μ'_C, μ_C, μ'_S, and μ_S be the corresponding optimal stable matchings. Then*

$$\mu_C \underset{C}{\geq} \mu'_C \text{ under } P \quad \text{and} \quad \mu'_C \underset{S}{\geq} \mu_C,$$

and

$$\mu'_S \underset{S}{\geq} \mu_S \quad \text{and} \quad \mu_S \underset{C}{\geq} \mu'_S \text{ under } P.$$

Symmetrical results are obtained if $\bar{P} \geq_S P$.

Proof: Suppose $\bar{P} \geq_C P$. Consider the marriage markets (S, C', P') and $(S, C', \bar{P}')$ related to (S, C, P) and $(S, C, \bar{P})$ respectively, where $P'(s) = \bar{P}'(s)$ for all s in S. Then $\bar{P}' \geq_{C'} P'$. If $\bar{P} \geq_S P$, consider the related marriage markets (S, C', P') and $(S, C', \bar{P}')$ and require that $\bar{P}' \geq_S P'$. Now apply Theorem 2.24.

Theorem 5.35 (Gale and Sotomayor). *Suppose C is contained in C^* and μ_S is the S-optimal matching for (S, C, P) and let μ_S^* be the S-optimal matching for (S, C^*, P^*), where P^* agrees with P on C. Then*

$$\mu_S^* \underset{S}{\geq} \mu_S \text{ under } P^* \quad \text{and} \quad \mu_S \underset{C}{\geq} \mu_S^*.$$

Symmetrical results are obtained if S is contained in S^*.

Proof: Suppose C is contained in C^*. Consider the marriage markets (S, C', P') and $(S, C^{*\prime}, P^{*\prime})$ related to (S, C, P) and (S, C^*, P^*) where P^* agrees with P' on C'. If S is contained in S^* it is enough to require that $P'(s) = P^{*\prime}(s)$ for all s in S. Now apply Theorem 2.25.

5.7   Stable matchings and the core

We now turn to the relationship between the set $S(P)$ of stable matchings and the core of the game. Following Definition 3.1, a matching μ' *dominates* another matching μ via a coalition A contained in $C \cup S$ if for all students s and colleges C in A,

> If $C' = \mu'(s)$ then $C' \in A$, and if $s' \in \mu'(C)$ then $s' \in A$; and
>
> $\mu'(s) >_s \mu(s)$, and $\mu'(C) >_C \mu(C)$.

Similarly, a matching μ' *weakly dominates* μ via a coalition A contained in $C \cup S$ if for all students s and colleges C in A,

> If $C' = \mu'(s)$ then $C' \in A$, and if $s' \in \mu'(C)$ then $s' \in A$; and
>
> $\mu'(s) \geq_s \mu(s)$, and $\mu'(C) \geq_C \mu(C)$

and

> $\mu'(s) >_s \mu(s)$ for some s in A, or
>
> $\mu'(C) >_C \mu(C)$ for some C in A.

That is, if μ' dominates μ via A, then every member of the effective coalition A strictly prefers μ' to μ, whereas if μ' weakly dominates μ via A, then every member of A likes μ' at least as much as μ, and at least one member of A strictly prefers μ' to μ.

The *core* of the game, $C(P)$, is the set of matchings that are not dominated by any other matching. The *core defined by weak domination*, $C_W(P)$, is the set of matchings that are not weakly dominated by any other matching. Since domination implies weak domination, $C_W(P)$ is contained in $C(P)$. When preferences are strict, the two cores coincide in the marriage model, but not in the college admissions model. However, when preferences are responsive, and when preferences over individuals are strict, the set of stable matchings coincides with the core defined by weak domination.

Proposition 5.36 (Roth). *When preferences over individuals are strict,* $S(P) = C_W(P)$.

Proof: If μ is not in $S(P)$, then μ is unstable via some student s and college C with $s >_C \sigma$ for some σ in $\mu(C)$. Then μ is weakly dominated via the coalition $C \cup s \cup \mu(C) - \sigma$ by any matching μ' with $\mu'(s) = C$ and $\mu'(C) = s \cup \mu(C) - \sigma$. In the other direction, if μ is not in $C_W(P)$, then μ is weakly dominated by some matching μ' via a coalition A, so some student or college in A prefers μ' to μ. (If μ is not individually rational, then it is also unstable, and we are done.) Suppose some C prefers μ' to μ. Then there must be some student s in $\mu'(C) - \mu(C)$ and some σ in $\mu(C) - \mu'(C)$ such that $s >_C \sigma$. (If not, then $\sigma \geq_C s$ for all s in $\mu'(C) - \mu(C)$ and σ in $\mu(C) - \mu'(C)$, which would imply $\mu(C) \geq_C \mu'(C)$, since C has responsive preferences.) So μ is unstable, since it is blocked by the pair (s, C). If some student s in A with $\mu'(s) = C$ prefers μ' to μ, then the fact that $\mu'(C) \geq_C \mu(C)$ similarly implies that there is a student s' (possibly different from s) in $\mu'(C) - \mu(C)$ and a σ in $\mu(C) - \mu'(C)$ such that $s' >_C \sigma$. Then μ is blocked by the pair (s', C).

So there may be unstable outcomes in the core, $C(P)$. Consider an example with one college, with quota 2, and three students, such that $P(C) = s_1, s_2, s_3$, and C is acceptable to each of the students. Then the only stable matching has $\mu(C) = \{s_1, s_2\}$. But the unstable matching ν with $\nu(C) = \{s_1, s_3\}$ is contained in $C(P)$, since it is not dominated via any coalition (since the coalition $\{C, s_2\}$ that prefers μ to ν is not effective for μ, and the members of the coalition $\{C, s_1, s_2\}$ that is effective for μ do not all prefer μ to ν).

Note that when preferences over individuals are not strict, the core defined by weak domination may be empty. Consider an example with two students and only one college, with a quota of one, who is acceptable to but indifferent between both students. Both individually rational matchings are weakly dominated, even though both are stable.

5.8   Guide to the literature

The "college admissions problem" is the name given by Gale and Shapley to a model of many-to-one matching that they considered together with the marriage model in their 1962 paper. In their model, only colleges' preferences over individuals, not over groups, are defined. They noted that the deferred acceptance algorithm for the marriage problem can be adapted to the case of many-to-one matching, and they showed that the existence of stable and optimal stable matchings as defined for the marriage problem carried over to the many-to-one case. The college admissions model presented in this chapter, in which colleges have preferences over groups of students as well as over individuals, was first presented in Roth (1985a).

For some time it was believed that problems of many-to-one matching were not different in any important way from problems of one-to-one matching, and attention was primarily directed to the marriage model. Because of this, a few incorrect assertions crept into the literature, growing out of the belief that results for the marriage model would automatically generalize to the case of many-to-one matching.

That this is not the case first became apparent in the study of the hospital–intern market reported in Roth (1984a). (Since that paper models an actual market, it was of course necessary to consider explicitly that each hospital program may employ more than one intern.) These modeling considerations led to the reformulation presented in Roth (1985a) and in this chapter. That paper introduced the notion of responsive preferences, and observed that certain theorems that were thought to carry over from the case of one-to-one matching do not. The model formulation and preliminary results presented in Sections 5.1.1 and 5.2 present the college admissions model as reformulated in Roth (1985a).

Section 5.4 concerning the hospital–intern market largely follows Roth (1984a), where Theorems 5.7, 5.8, 5.11, and 5.12 are presented. Theorem 5.11, which shows that the market with couples may have no stable outcomes, was also independently proved by Sotomayor in an unpublished note. Theorems 5.10 and 5.14, which show that Theorems 2.27, 4.7, and 4.10 do not generalize from the case of the marriage market, are from Roth (1985a), as are Theorems 5.16 and 5.18. Theorem 5.13 is from Roth (1986).

Theorems 5.11 and 5.14 show that certain things may not always happen - there may not always be a stable matching in a market with couples, and it may not always be a best response for a hospital to state its true preferences – but they do not give any indication of how often this may be the case, or how hard it may be to determine just what the situation is. However some progress has been made on these questions. Regarding Theorem 5.11, Ronn (1986, 1987) has shown that the problem of determining if a market with couples has any stable matchings is computationally complex. In particular, it is "*NP* complete," which is a measure of the time required for computation. When all preferences over individuals are strict, Ronn shows that the problem is also "logspace *P*-hard," which is a measure of the computer memory required. (See Garey and Johnson, 1979.) So for large markets with couples, it may not always be practical to find a stable matching even when one exists. And regarding Theorem 5.14, some preliminary work by Wood (1984) suggests that the example used in the proof, in which some hospital can profitably misrepresent its preferences, may be neither perverse nor uncommon.

Section 5.5 concerning the medical markets in the United Kingdom follows Roth (1989b). In addition to the empirical work reported in Roth (1984a, 1989b) concerning the institutions by which physicians get their first jobs in the United States and the United Kingdom, another matching mechanism that has been studied in detail is the "preferential bidding system" used by American college sororities to match new members to sororities (Mongell 1988, and Mongell and Roth 1989). Some preliminary empirical work concerning various entry level professional labor markets that do not use centralized market mechanisms is also under way. A laboratory experiment concerned with decentralized markets is reported by Sondak and Bazerman (1988).

Section 5.6 closely follows the paper of Roth and Sotomayor (1989). Lemma 5.25 and Theorems 5.26 through Corollary 5.28 are from that paper. The proof of Lemma 5.25 given here is shorter than the original proof of Roth and Sotomayor, thanks to a suggestion by David Gale. Theorems 5.29 and 5.31, which generalize results already known for the marriage model, may have been known for some formulations of the college admissions model as well, but as far as we know are first formally proved here. Theorems 5.34 and 5.35 are from Gale and Sotomayor (1985a).

Proposition 5.36, concerning the relationship between the set of stable matchings and the core defined by weak domination, was proved by Roth (1985b) under the assumption that colleges' preferences over groups are strict.

Finally, we remark that the study of the hospital–intern market allows some interesting commentary on the history of ideas. The NIMP algorithm, which is a stable matching mechanism (Theorem 5.7), was first used in 1951, a full decade before the paper of Gale and Shapley, and contemporaneously (but quite independently) with the definition in the game-theoretic literature of the core of a game by Gillies (1953a, b) and Shapley

(1953a). (See Stalnaker (1953) for an early account of the NIMP algorithm.) The NIMP algorithm replaced the earlier trial run algorithm because that earlier algorithm did not make it a dominant strategy for agents to state their true preferences. Although it turns out (Theorems 4.4 and 5.14) that neither does the NIMP algorithm, this kind of consideration did not enter the theoretical literature of economics until much later; see, for example, the references in Green and Laffont (1979). So a number of ideas that have subsequently become important in economic theory came up in a practical way in the organization of this market. And in a case of art imitating life imitating art, some of the work of Gale and Shapley (1962) and Roth (1984a) has begun to be reflected in the medical literature concerned with these markets. For example, the NIMP algorithm is now described to participants in terms of the deferred acceptance procedure (in this respect compare the current description (NRMP Directory 1987) with earlier ones (e.g., NIRMP Directory 1979)).

This history gives us an opportunity to reflect briefly on the role of scientific "firsts," a subject about which there sometimes seems to be unnecessary confusion. That the 1951 NIMP algorithm is a stable matching mechanism, produced by a committee concerned with a practical problem, is a remarkable and noteworthy achievement. But the subsequent discovery and appreciation of this achievement does not diminish the contribution of Gale and Shapley (1962), who independently developed a closely related mechanism. Perhaps the point can be made clear by noting that Columbus is viewed as the discoverer of America, even though every school child knows that the Americas were inhabited when he arrived, and that he was not even the first to have made a round trip, having been preceded by Vikings and perhaps by others. What is important about Columbus's discovery of America is not that it was the first, but that it was the *last*. After Columbus, America was never lost again; no subsequent explorers can claim its discovery. But to make a somewhat different point, just as the importance of what Columbus discovered is in large measure due to what was subsequently learned about and accomplished in the Americas, so it is with scientific theories and discoveries, whose importance derives from what they allow us to understand and to predict.

CHAPTER 6

Discrete models with money, and more complex preferences

This chapter looks at models of many-to-one matching between firms and workers that generalize the college admissions model in two important ways, by allowing firms to have a larger class of preferences over groups of workers, and by explicitly putting money into the model, so salaries are determined as part of the outcome of the game, rather than specified in the model as part of the job description. Of course, when we look at a model that does both these things together, we will also have to specify how the preferences of firms and workers deal with different combinations of job assignments and wages. Section 6.2 considers a version of a model proposed by Kelso and Crawford, with both complex preferences and negotiated wages. (We model wages here as a discrete variable, which is a natural modeling assumption since, for example, contracts cannot specify wages more closely than to the nearest penny. The next chapters model wages as a continuous variable, which also has some advantages.) But first we construct a simpler model in which we can examine complex preferences while continuing to treat salaries as an implicit part of the job description. Throughout this chapter we continue to make the simplifying assumption that workers are indifferent to which other workers are employed by the same firm.

6.1 The college admissions model with "substitutable" preferences

We noted in Chapter 5 that in the college admissions model, when colleges' preferences over groups of students are responsive to their preferences over individual students, the set of stable matchings is nonempty, but that it may be empty when preferences are not responsive (as in Example 2.7). In this section, we consider a weaker condition than responsiveness that is nevertheless sufficient to preserve not only the nonemptiness of the set of stable matchings, but also many of the properties noted in Chapter 5. The basic idea is that the set of stable matchings will continue to be nonempty so long as firms' preferences for groups of workers are such that the firms regard individual workers more as substitutes for each other than as complements. It turns out that this condition will make the deferred acceptance algorithm operate just as it did when we assumed colleges have responsive preferences.

Let the two sets of agents be n firms $F = \{F_1, \ldots, F_n\}$, and m workers $W = \{w_1, \ldots, w_m\}$. For simplicity assume all firms have the same quota, equal to m, so each firm could in principle hire all the workers. This will allow us to describe matchings a little more simply, since it will not be necessary to keep track of each firm's quota by saying, for example, that a firm that doesn't employ any workers is matched to m copies of itself.

Definition 6.1.   *A matching μ is a function from the set $F \cup W$ into the set of all subsets of $F \cup W$ such that*

1. $|\mu(w)| = 1$ *for every worker w and $\mu(w) = w$ if $\mu(w) \notin F$;*
2. $|\mu(F)| \leq m$ *for every firm F ($\mu(F) = \phi$ if F isn't matched to any workers);*
3. $\mu(w) = F$ *if and only if w is in $\mu(F)$.*

Workers have preferences over individual firms, just as in the college admissions problem, and firms have preferences over subsets of W. For simplicity we are going to assume that all preferences are strict. So a worker w's preferences can be represented by a list of acceptable firms, for example, $P(w) = F_i, F_j, F_k, w$; and a firm's preferences by a list of acceptable subsets of workers, for example, $P^\#(F) = S_1, S_2, \ldots, S_k, \phi$; where each S_i is a subset of W. Each agent compares different matchings by comparing his or her (its) own assignment at those matchings. The preferences of all the agents will be denoted by $P = (P^\#(F_1), \ldots, P^\#(F_n), P(w_1), \ldots, P(w_m))$. Keep in mind that the preferences of the firms are preferences over *sets* of employees.

Faced with a set S of workers, each firm F can determine which subset of S it would most prefer to hire. We will call this F's choice from S, and denote it by $Ch_F(S)$. That is, for any subset S of W, F's *choice set* is $Ch_F(S) = S'$ such that S' is contained in S and $S' \geq_F S''$ for all S'' contained in S. Since we have assumed that preferences are strict, there is always a single set S' that F most prefers to hire, out of any set S of available workers. (Of course S' could equal S, or it could be empty.) We can now formalize what we mean when we say that we are going to require that firms regard workers as substitutes rather than complements.

Definition 6.2. *A firm F's preferences over sets of workers has the property of substitutability if, for any set S that contains workers w and w', if w is in $Ch_F(S)$ then w is in $Ch_F(S-w')$.*

That is, if F has "substitutable" preferences, then if its preferred set of employees from S includes w, so will its preferred set of employees from any subset of S that still includes w. (By repeated application, if $w \in Ch_F(S)$ then for any S' contained in S with $w \in S'$, $w \in Ch_F(S')$.) This is the sense in which the firm regards worker w and the other workers in $Ch_F(S)$ more as substitutes than complements: It continues to want to employ w even if some of the other workers become unavailable.

So substitutability rules out the possibility that firms regard workers as complements, as might be the case of an American football team, for example, that wanted to employ a player who could throw long passes and one who could catch them, but if only one of them were available would prefer to hire a different player entirely. Note that in Example 2.7 this kind of complementarity appears in the preferences of firm F_1, since its preferences are such that $Ch(\{w_1, w_3\}) = \{w_1, w_3\}$, but $Ch(\{w_3\}) = \phi$. That is, F_1 would like to hire both w_1 and w_3, but if w_1 isn't available then it is no longer interested in w_3. We will see that the emptiness of the set of stable matchings in Example 2.7 is related to this complementarity, that is, to the fact that firms in that example do not have substitutable preferences.

Note that responsive preferences have the substitutability property. In the college admissions model, the choice set from any set of students of a college with quota q is either the q most preferred acceptable students in the set, or all the acceptable students in the set, whichever is the smaller number.

A matching μ is *blocked* by an individual worker w if $w >_w \mu(w)$, and by an individual firm F if $\mu(F) \neq Ch_F(\mu(F))$. Note that μ may be blocked by an individual firm F without being individually irrational, since it might still be that $\mu(F) >_F \phi$. Our definition of blocking recognizes that F may fire some workers in $\mu(F)$ if it chooses, without affecting other members of $\mu(F)$.

Similarly, μ is blocked by a worker–firm pair (w, F) if w and F aren't matched at μ but would both prefer if F hired w; that is, if $\mu(w) \neq F$ and if $F >_w \mu(w)$ and $w \in Ch_F(\mu(F) \cup w)$. If the firms have responsive preferences, this definition is equivalent to the one we used for the college admissions model. We go about defining stable matchings in the same way also.

Definition 6.3. *A matching μ is **stable** if it is not blocked by any individual agent or any worker-firm pair.*

Since blocking is now defined in terms of firms' preferences over sets of workers, this definition of stability has a slightly different meaning than the same definition in the previous chapter. Nevertheless, it is still a definition of *pairwise* stability, since the largest coalitions it considers are worker–firm pairs. So we still have to consider whether something is missed by not considering larger coalitions. It turns out that pairwise stability is still sufficient – as when preferences are responsive, we can show that the set $S(P)$ of stable matchings equals the set of group stable matchings. In fact, the definitions of the core, $C(P)$, and the core defined by weak domination, $C_W(P)$, are identical to those presented in Section 5.7, and as in the case of responsive preferences it is not hard to show that $C(P)$ contains $S(P)$, which in turn contains $C_W(P)$. We prove below the extension of Proposition 5.36 to the present model.

Proposition 6.4. *When firms have substitutable preferences (and all preferences are strict)* $S(P) = C_W(P)$.

Proof: If μ is not in $S(P)$, then it is blocked by some individual or some pair (w, F). If it is blocked by F alone, then it is weakly dominated by a matching μ' such that $\mu'(F) = Ch_F(\mu(F))$, via the coalition $A = \{F \cup \mu'(F)\}$. If μ is blocked by w alone, then it is weakly dominated by μ' such that $\mu'(w) = w$, via the coalition $A = \{w\}$. And if μ is blocked by the pair (w, F), then it is weakly dominated by μ' such that $\mu'(F) = Ch_F(\mu(F) \cup w) \neq \mu(F)$ via the coalition $A = F \cup \mu'(F)$. (If $w' \in \mu'(F)$ then either $w' = w$ or $w' \in \mu(F)$, so either $F >_{w'} \mu(w')$ or $\mu(w') = \mu'(w')$.) So $S(P)$ contains $C_W(P)$.

In the other direction, if μ is weakly dominated by μ' via a coalition A, then there is some agent in A who prefers μ' to μ. We may assume μ is not blocked by any individual, since if it is it is unstable and we are done. Suppose some worker w in A prefers μ' to μ, and let $F = \mu'(w) \neq \mu(w)$. Then $\mu'(F) >_F \mu(F)$, since preferences are strict. So there is a worker w' in $Ch_F(\mu'(F) \cup \mu(F))$ such that w' is not in $\mu(F)$ (otherwise μ is blocked by F alone). So $w' \in \mu'(F)$ and hence $w' \in A$ and $F = \mu'(w') \neq \mu(w')$ so $\mu'(w') >_{w'} \mu(w')$ by strict preferences. By substitutability, w' is in $Ch_F(\mu(F) \cup w')$, so μ is blocked by (w', F) and is unstable. If it is some firm F in A that prefers μ', then there is some worker w in $Ch_F(\mu'(F) \cup \mu(F))$ such that w is not in $\mu(F)$, since otherwise F blocks μ alone. So w is in A, and prefers $F = \mu'(w)$ to $\mu(w) \neq F$ since preferences are strict. But $w \in Ch_F(\mu(F) \cup w)$ by substitutability, so (w, F) blocks μ. So $S(P)$ is contained in $C_W(P)$.

We can now state our main result.

Theorem 6.5.   *When firms have substitutable preferences, the set of stable matchings is always nonempty.*

Before proving this, an example will help clarify things.

Example 6.6: An example in which firms have substitutable
(but nonresponsive) preferences
There are two firms and three workers, with preferences as follows.

$$P^{\#}(F_1) = \{w_1, w_2\}, \{w_1, w_3\}, \{w_2, w_3\}, \{w_3\}, \{w_2\}, \{w_1\}$$
$$P^{\#}(F_2) = \{w_3\}$$
$$P(w_1) = F_1, F_2$$
$$P(w_2) = F_1, F_2$$
$$P(w_3) = F_1, F_2$$

Note that

$$\mu = \begin{matrix} F_1 & F_2 \\ \{w_1, w_2\} & \{w_3\} \end{matrix}$$

is the unique stable matching.

If we look just at single workers, we see that F_1 prefers w_3 to w_2 to w_1. But $P^{\#}(F_1)$ is not responsive to these preferences over single workers, since $\{w_1, w_2\} >_{F_1} \{w_1, w_3\}$ even though w_3 alone is preferred to w_2 alone. But the preferences are substitutable.

Recall the discussion in Section 5.1.1, and observe once again that the class of college admissions problems of which this is an example would not be well-defined games if we specified only the preferences of firms over individuals. Indeed, if we defined stability only in terms of preferences over individuals, the matching μ would be unstable with respect to the pair (F_1, w_3) since w_3 prefers F_1 to F_2 and F_1 prefers w_3 (by himself) to w_2 (by himself). But μ is not unstable in this example because F_1 prefers $\{w_1, w_2\}$ to any of $\{w_1, w_3\}$, $\{w_2, w_3\}$, or $\{w_3\}$.

We now turn to the proof of Theorem 6.5. The proof will be by means of the deferred acceptance algorithm with firms proposing, which needs to be only minimally adapted to the current model, as follows. In the first step, each firm proposes to its most preferred set of workers, and each worker rejects all but the most preferred acceptable firm that proposes to him or her. In each subsequent step, each firm proposes to its most preferred set of workers that includes all of those workers whom it previously proposed to and who have not yet rejected it, but does not include any workers who have previously rejected it. Each worker rejects all but the most preferred acceptable firm that has proposed so far. The algorithm stops after any step in which there are no rejections, at which point each firm is matched to the set of workers to which it has issued proposals that have not been rejected.

Proof of Theorem 6.5:   We will show that the matching μ produced by the algorithm just described is stable. The key observation is that because firms have substitutable preferences, no firm ever regrets that it must continue to offer employment at subsequent steps of the algorithm to workers who have not rejected its earlier offers. That is, at every step in the algorithm, each firm is proposing to its most preferred set of workers that does not contain any workers who have previously rejected it. So consider a firm F and worker w such that $w \in Ch_F(\mu(F) \cup w)$. At some step of the algorithm, F proposed to w and was subsequently rejected, so w prefers $\mu(w)$ to F, and μ is not blocked by the pair (w, F). Since w and F were arbitrary, and since μ is not blocked by any individual, μ is stable.

In fact, the matching μ is the firm-optimal stable matching, which gives each firm its most preferred set of achievable workers.

Theorem 6.7.   *When firms have substitutable preferences (and preferences are strict) the deferred acceptance algorithm with firms proposing produces a firm-optimal stable matching.*

We leave to the reader the proof, which can be constructed following the proof of Theorem 2.12, by showing that no firm is ever rejected by an achievable worker in the deferred acceptance algorithm with firms proposing.

Since this model is not symmetric between firms and workers, it is not immediately apparent that the deferred acceptance algorithm with workers proposing will have an analogous result, but it does. When the workers propose, the description of the algorithm needs even less modification, since workers propose to firms in order of preference, and a firm rejects at any step all those workers who are not in the firm's choice set from those proposals it has not yet rejected. We can state the following result.

Theorem 6.8.   *When firms have substitutable preferences (and preferences are strict) the deferred acceptance algorithm with workers proposing produces a worker-optimal stable matching.*

The key observation for the proof, which we also leave to the reader, is that because firms have substitutable preferences, no firm ever regrets that it rejected a worker at an earlier step when it sees who proposed at the current step.

We close this section with an example that shows that these results cannot be generalized in a straightforward way to the symmetric case of many-to-many matching in which workers may take multiple jobs, even when both sides have substitutable preferences. The reason is not that the analogously defined pairwise stable matchings do not have similar properties in such a model, but that pairwise stable matchings are no longer group stable.

Example 6.9: Many-to-many matching with substitutable
preferences (Blair)
There are three firms and three workers (each with a quota of two or three), with preferences as follows.

$$P^\#(F_1) = \{w_1, w_2\}, \{w_2, w_3\}, \{w_1\}, \{w_2\}, \{w_3\}$$
$$P^\#(F_2) = \{w_2, w_3\}, \{w_1, w_3\}, \{w_2\}, \{w_1\}, \{w_3\}$$
$$P^\#(F_3) = \{w_1, w_3\}, \{w_1, w_2\}, \{w_3\}, \{w_1\}, \{w_2\}$$
$$P^\#(w_1) = \{F_1, F_2\}, \{F_2, F_3\}, \{F_1\}, \{F_2\}, \{F_3\}$$
$$P^\#(w_2) = \{F_2, F_3\}, \{F_1, F_3\}, \{F_2\}, \{F_1\}, \{F_3\}$$
$$P^\#(w_3) = \{F_1, F_3\}, \{F_1, F_2\}, \{F_3\}, \{F_1\}, \{F_2\}$$

The workers and firms play completely symmetric roles in one another's preferences, and there is a unique (pairwise) stable matching μ given by

$$\mu(F_i) = \{w_i\} \quad \text{for } i = 1, 2, 3.$$

However μ is dominated via the coalition of all agents by the matching μ' given by

$$\mu'(F_1) = \{w_2, w_3\}, \qquad \mu'(F_2) = \{w_1, w_3\}, \qquad \mu'(F_3) = \{w_1, w_2\};$$
$$\mu'(w_1) = \{F_2, F_3\}, \qquad \mu'(w_2) = \{F_1, F_3\}, \qquad \mu'(w_3) = \{F_1, F_2\}.$$

The matching μ' is in the core in this example (defined in the conventional way) but it is not pairwise stable.

6.2   A model with money and complex preferences

This section presents a version of the matching model of Kelso and Crawford in which the idea of substitutable preferences was introduced. In this model, firms and workers have preferences both for whom they are matched with, and what the salaries are. So an outcome of the market will consist not only of a matching but also of a division of the gains from each matching into profit for the firms and salaries for the workers.

Let the agents be a set of m workers W indexed by $i = 1, \ldots, m$ and a set of n firms F indexed by $j = 1, \ldots, n$. Each firm can hire as many workers as it wishes (i.e., the quota of each firm is m) and each worker can work at only one firm.

The preferences of the workers will be represented by utility functions that depend on who they work for, and at what salary. So the utility to worker i of working for firm j at salary s_i is given by a strictly increasing function $u_{ij}(s_i)$. (Note that we continue to assume workers are unconcerned with who their coworkers are.) For each worker i there is a vector $\sigma^i \equiv (\sigma_{i1}, \ldots, \sigma_{in})$ where σ_{ij} represents the lowest salary that worker i would accept to work for firm j. That is, the utility to worker i is the same whether he or she works for j at the salary σ_{ij} or is unemployed at zero salary. Denote by $u_{i0}(0)$ the utility to worker i of being unemployed at zero salary.

For each firm j and each subset C of workers, there is a nonnegative number $Y^j(C)$ representing the amount of income that will accrue to the firm when its employees are precisely the set C of workers. The assumptions about these numbers are that for all workers i and firms j,

   (i) $Y^j(\phi) = 0$; and
   (ii) $Y^j(C \cup \{i\}) - Y^j(C) \geq \sigma_{ij}$ for any set C of workers that does not contain worker i.

The first condition says that each firm must have some employees to produce anything, and the second says that the marginal contribution of each worker to each firm is never less than the salary that would make the worker indifferent between working or being unemployed. This second assumption needs some comment, since it is not as strong as it might seem. Indeed, condition (ii) can be regarded as a modeling convention, since if a worker's marginal product were less than the minimum wage that would make him or her indifferent to unemployment, the firm could agree to let the worker do nothing for a zero salary. In this model, such a worker will appear employed at salary $\sigma_{ij} = 0$, rather than unemployed.

So in this model there will be no unemployed workers (at least not in the core), since hiring one more worker i at salary σ_{ij} never hurts firm j. Thus a *matching* can be taken to be a set of disjoint partnerships of form $\{j, C^j\}$ or $\{j\}$, where $\{j, C^j\}$ denotes that firm j employs the set C^j of workers, and $\{j\}$ denotes that firm j employs no workers. An *outcome* for this model consists of a matching μ and, for each partnership $\{j, C^j\}$ in μ, an allocation of income $Y^j(C^j)$ into π_j (profit) and $\{s_i, i \in C^j\}$ (salaries) such that $Y^j(C^j) = \pi_j + \sum_{i \in C^j} s_i$. If a firm j is unmatched, then

$\pi_j = 0$. We will denote an outcome by a triple (μ, π, s), where π is the vector of profits for each firm j, and s is the vector of salaries paid to each worker i by firm $\mu(i)$. An outcome (μ, π, s) is *individually rational* if $s_i \geq \sigma_{i\mu(i)}$ for each worker i, and $\pi_j \geq 0$ for each firm j.

As indicated earlier, we will model salaries as discrete variables, the idea being that there is some smallest unit (e.g., pennies per hour or dollars per year) beyond which salaries cannot be further divided. To facilitate this, we assume that all salaries s_i and incomes and profits $Y^j(C^j)$ and π_j are stated in these units, so that they will all take on only integer values.

An individually rational outcome (μ, π, s) will be called a core allocation unless there is a firm j, a subset of workers C, and a vector r of integer salaries r_i, for all workers i in C, such that

$$\pi_j < Y^j(C) - \sum_{i \in C} r_i \quad \text{and}$$

$$u_{i\mu(i)}(s_i) < u_{ij}(r_i)$$

for all workers i in C.

If these two inequalities are satisfied for some (j, C, r), then the outcome (μ, π, s) is *blocked* by (j, C, r). The first inequality says that firm j can make a higher profit by employing the set of workers C, with salaries r_i, than by employing the set of workers matched to it by μ at the current salaries s_j. The second inequality says that every worker i in the set C prefers to work for firm j at salary r_i rather than continue to work for firm $\mu(i)$ at salary s_i.

If no further restrictions are imposed on the model the core may be empty.

Example 6.10: An example in which the core is empty
Consider a market that consists of firms j and k and workers 1 and 2. The amounts of income $Y^j(C)$ and $Y^k(C)$ for all possible subsets of workers are given by

$$Y^j(\{1\}) = 4 \qquad Y^k(\{1\}) = 8$$
$$Y^j(\{2\}) = 1 \qquad Y^k(\{2\}) = 5$$
$$Y^j(\{1, 2\}) = 10 \qquad Y^k(\{1, 2\}) = 9.$$

The two workers are indifferent between the two firms and care only about salary, so $u_{ij}(s_i) = s_i$ and $u_{ik}(s_i) = s_i$, and $\sigma_{ij} = \sigma_{ik} = 0$ for both workers $i = 1, 2$.

The only matchings at which neither worker is unemployed are

$$\mu_1 = \begin{pmatrix} j & k \\ \{1,2\} & - \end{pmatrix} \quad \mu_2 = \begin{pmatrix} j & k \\ - & \{1,2\} \end{pmatrix} \quad \mu_3 = \begin{pmatrix} j & k \\ \{1\} & \{2\} \end{pmatrix} \quad \mu_4 = \begin{pmatrix} j & k \\ \{2\} & \{1\} \end{pmatrix}.$$

We will show there is no allocation (π, s) such that (μ_i, π, s) is a core outcome for any of these μ_i. (It is readily seen that no matching with an unemployed worker can be in the core.)

Consider $s = \{s_1, s_2\}$, $\pi = \{\pi_j = 10 - \{s_1 + s_2\}, \pi_k = 0\}$ and the matching μ_1. This outcome is blocked by $(k, \{1\}, r_1 = s_1 + 1)$ or by $(k, \{2\}, r_2 = s_2 + 1)$, since otherwise, $0 = \pi_k \geq 8 - (s_1 + 1)$ and $0 = \pi_k \geq 5 - (s_2 + 1)$, and hence $s_1 + s_2 \geq 11$, which contradicts that $\pi_j \geq 0$.

Now consider the outcome $(\mu_2, \{\pi_j = 0, \pi_k = 9 - (s_1 + s_2)\}, \{s_1, s_2\})$. In order for this outcome not to be blocked by $(k, \{1\}, s_1 + 1)$ or by $(k, \{2\}, s_2 + 1)$ we must have $9 - (s_1 + s_2) = \pi_k \geq 8 - (s_1 + 1)$ and $9 - (s_1 + s_2) = \pi_k \geq 5 - (s_2 + 1)$, and so $s_1 + s_2 \leq 7$. But then the triple $(j, \{1, 2\}, r_1 = s_1 + 1, r_2 = s_2 + 1)$ blocks (μ_2, π, s) since otherwise we would have that $0 = \pi_j \geq 10 - (s_1 + s_2 + 2)$ and then $s_1 + s_2 \geq 8$.

The outcome $(\mu_3, \{\pi_j = 4 - s_1, \pi_k = 5 - s_2\}, \{s_1, s_2\})$ is blocked by $(k, \{1\}, r_1 = s_1 + 1)$ or $(j, \{1, 2\}, r_1 = s_1 + 1, r_2 = s_2 + 1)$, since otherwise $5 - s_2 = \pi_k \geq 8 - (s_1 + 1)$ and $4 - s_1 = \pi_j \geq 10 - (s_1 + s_2 + 2)$, and so $s_1 \geq 6$. But this contradicts the fact that $\pi_j \geq 0$. It is left to the reader to verify that the outcome $(\mu_4, \{\pi_j = 1 - s_2, \pi_k = 8 - s_1\}, \{s_1, s_2\})$ is blocked by $\{k, \{2\}, r_2 = s_2 + 1\}$ or $\{j, \{1, 2\}, r_1 = s_1 + 1, r_2 = s_2 + 2\}$.

Notice that in this example the technology of firm j, as reflected in the function Y^j, exhibits a kind of increasing returns, since with workers 1 and 2 the firm generates more income than the sum of the income from each worker separately. These increasing returns induce some complementarity into the preferences of firm j, since, for example, it is willing to employ both workers at salaries of $s_1 = s_2 = 4$, but if worker 1 were to demand a salary of 7, firm j would be unwilling to employ either worker. That is, firm j is willing to employ worker 2 at a salary of 4 when worker 1 is also employed at a salary of 4, but not when worker 1 is unavailable (or too high-priced).

In order to rule out this kind of complementarity in the preferences of the firms, Kelso and Crawford proposed a condition on the functions Y^j. To state it, we need to introduce some notation.

Let $s = (s_1, \ldots, s_m)$ be a vector of salaries. Let $M^j(s)$ denote the set of solutions to the problem: "Choose C^j to maximize $(Y^j(C) - \Sigma_{i \in C} s_i)$, for all possible sets C of workers."

If we interpret s as the vector of salaries that firm j would have to pay to attract each worker i, then the firm's problem stated above is to choose (one of) the most profitable sets of workers. Now consider another vector of salaries $\bar{s}$. Let $T^j(C^j) = \{i \mid i \in C^j \text{ and } \bar{s}_i = s_i\}$. That is, $T^j(C^j)$ is the set of workers in (one of) firm j's choice set(s) at salaries s whose salary demands are unchanged at $\bar{s}$.

The *gross substitutes* assumption of Kelso and Crawford is that

(iii)   for every firm j, if $C^j \in M^j(s)$ and $\bar{s} \geq s$, then there exists $\bar{C}^j \in M^j(\bar{s})$ such that $T^j(C^j)$ is contained in $\bar{C}^j$.

That is, if a worker i is in the choice set of firm j when the salaries the firm must pay to hire each worker are given by s, then the firm will still want to hire worker i if the salaries demanded by *other* workers rise, but worker i's salary demand does not. Note that this model requires a more complicated statement of "substitutability" than was needed in the simpler model of Section 6.1, for two reasons. First, in this model preferences must take salaries into account. Second, preferences are not assumed to be strict, so there may be multiple sets of workers that solve the choice problem facing a firm for any vector of salaries demanded.

In Example 6.10 the gross substitutes condition is not satisfied by firm j. Consider the two vectors of salaries $s = (4, 4)$ and $\bar{s} = (7, 4)$: The unique preferred set of workers for firm j at salaries s is $\{1, 2\}$ and at salaries $\bar{s}$ is ϕ.

We can now state the following.

Theorem 6.11 (Kelso and Crawford).   *When the gross substitutes condition applies to all firms, the core is nonempty.*

To prove Theorem 6.11, Kelso and Crawford (1982) proposed the following variant of the deferred acceptance algorithm. The algorithm proceeds by the following rules. We quote:

R(1) - Firms begin facing a set of permitted salaries $s_{ij}(0) = \sigma_{ij}$. Permitted salaries at round t, $s_{ij}(t)$, remain constant, except as noted below. In round zero, each firm makes offers to all workers; this is costless by condition (ii).

R(2) - On each round, each firm makes offers to the members of one of its favorite sets of workers, given the schedule of permitted salaries

$$s^j(t) = [s_{1j}(t), \ldots, s_{mj}(t)].$$

That is, firm j makes offers to the members of $C^j(s^j(t))$, where $C^j(s^j(t))$ maximizes $Y^j(C) - \sum_{i \in C} s_{ij}(t)$. Firms may break ties between sets of workers however they like, with the following exception: any offer made by firm j in round $t-1$ that was not rejected must be repeated in round t. By the gross substitutes condition, the firm sacrifices no profits in doing this, since (by R4) other workers' permitted salaries cannot have fallen, and the salary of a worker who did not reject an offer remains constant.

R(3) - Each worker who receives one or more offers rejects all but his or her favorite (taking salaries into account), which he or she tentatively accepts. Workers may break ties at any time however they like.

R(4) – Offers not rejected in previous periods remain in force. If worker i rejected an offer from firm j in round $t-1$, $s_{ij}(t) = s_{ij}(t-1) + 1$; otherwise $s_{ij}(t) = s_{ij}(t-1)$. Firms continue to make offers to their favorite sets of workers, taking into account their permitted salaries.

R(5) – The process stops when no rejections are issued in some period. Workers then accept the offers that remain in force from the firms they have not rejected.

Proof of Theorem 6.11: We will show that the outcome produced by the algorithm is in the core.

It is easy to see that the process stops after a finite number of rounds. If $s = \{s_1, \ldots, s_m\}$ is the set of salaries finally accepted by the workers and μ is the matching when the process stops, the set $\pi = \{\pi_1, \ldots, \pi_n\}$ of profits for the firms is given by $\pi_j = Y^j(C^j) - \sum_{i \in C^j} s_i$, if $\{C^j, j\}$ belongs to the matching μ.

It is immediate from R1 and R4 that $s_i \geq \sigma_{ij}$, for all $i = 1, \ldots, m$, if i and j are matched by μ. It is also immediate from R2 and the fact that the firm is not required to hire workers that $\pi_j \geq 0$ for all $j = 1, \ldots, n$.

When the algorithm stops, every worker is holding exactly one offer. Furthermore the set of workers C^j assigned by μ to j gives to j the maximum net profit it could get among all possible subsets of workers at salaries $s_{ij}(t^*)$, where t^* is the round at which the algorithm stopped. That is,

$$\pi_j \geq Y^j(C) - \sum_{i \in C} s_{ij}(t^*), \tag{1}$$

for all subsets of workers C.

Then if some (j, C, r) blocked (μ, π, s), where r is a set of integer salaries, we should have

$$u_{ij}(r_i) > u_{i\mu(i)}(s_i) \quad \text{for all } i \text{ in } C \tag{2}$$

and

$$Y^j(C) - \sum_{i \in C} r_i > \pi_j \tag{3}$$

By (2) and R3, worker i must never have received an offer from firm j at salary r_i or greater, for all i in C. So $s_{ij}(t^*) \leq r_i$ for all i in C. But then

$$\pi_j < Y^j(C) - \sum_{i \in C} r_i \leq Y^j(C) - \sum_{i \in C} s_{ij}(t^*),$$

which contradicts (1).

Kelso and Crawford went on to observe that when all firms and workers have strict preferences, the outcome produced by this algorithm is the firm-optimal core outcome. They also showed that if a new firm is added to the market, each worker does at least as well at the firm-optimal outcome of the new market as at the corresponding outcome of the original market, and if a worker is removed from the market, then each firm is at least as well off at the firm-optimal outcome of the original market as at the corresponding outcome of the new market. This latter result is thus a precursor of Theorems 2.25 and 5.35 for the marriage and college admissions models, as well as of similar results for the models of one-to-one matching with money to be studied in Chapters 8 and 9.

It is the assumption that preferences obey the gross substitutes condition that allows the potentially complex preferences over groups introduced in this model to be treated like the simpler preferences we study in the other models in this book. It is therefore important to note that this is quite a strong assumption. For example, aside from ruling out that firms may have technologies with increasing returns of the sort exhibited in Example 6.10, the assumption also rules out simple budget constraints. We might expect that firms would face budget constraints if it takes a long time to produce the product that they sell, so that workers must be hired (and paid) substantially in advance of the time that the product of their labor can be sold. In such a case, there might be a practical limit on how much a firm could borrow in anticipation, so that salaries of workers would have to be paid out of an existing stream of funds. The following example shows that even when the firm's technology obeys the gross substitutes assumption, the introduction of budget constraints may introduce complementarities that will cause the core to be empty.

Example 6.12: An example with budget constraints and an empty core (Mongell and Roth)
The market consists of two firms, j and k, and three workers, 1, 2, and 3, as follows.

Firm j	Firm k
$Y^j(\{1\}) = 700$	$Y^k(\{1\}) = 600$
$Y^j(\{2\}) = 1400$	$Y^k(\{2\}) = 1500$
$Y^j(\{3\}) = 800$	$Y^k(\{3\}) = 1100$
$Y^j(\{1, 2\}) = 2100$	$Y^k(\{1, 2\}) = 2100$
$Y^j(\{1, 3\}) = 1500$	$Y^k(\{1, 3\}) = 1700$
$Y^j(\{2, 3\}) = 2200$	$Y^k(\{2, 3\}) = 2600$
$Y^j(\{1, 2, 3\}) = 2900$	$Y^k(\{1, 2, 3\}) = 3200$
$Y^j(\{\emptyset\}) = 0$	$Y^k(\{\emptyset\}) = 0$
$\sigma_{1j} = 400$	$\sigma_{1k} = 300$
$\sigma_{2j} = 700$	$\sigma_{2k} = 1000$
$\sigma_{3j} = 400$	$\sigma_{3k} = 700$

Note Y^j and Y^k are separable (the product of two workers is just the sum of their separate products) so the gross substitutes condition holds for both firms.

Now impose a budget constraint on firms j and k, respectively; $\sum s_i \le$ $B^j = 440$ for $i \in C^j$ and $\sum s_i \le B^k = 1075$ for $i \in C^k$. An outcome must meet the requirement that the sum of the salaries of workers employed by a firm does not exceed the firm's budget. The gross substitutes condition applied to the preferences with this constraint no longer holds for firm k. Consider the two salary vectors $s = (300, 1000, 700)$ and $\bar{s} = (380, 1000, 700)$. When the salaries are s, the unique preferred set of workers for firm k (subject to its budget constraint) is $\{1, 3\}$. But when the salaries demanded are $\bar{s}$, firm k cannot afford to hire the set of workers $\{1, 3\}$ and chooses the set $\{2\}$. That is, even though worker 3's salary demand does not change, the firm only chooses to employ him or her if worker 1 can be hired at the lower salary.

The core in this example is empty. One way to organize the somewhat tedious computations needed to check this is first to observe that there exists no core outcome at which worker 2 is employed, and then to check that there is also no core outcome at which worker 2 is unemployed.

Note that budget constraints therefore have quite a different effect than the kind of "gender budget constraint" we encountered in Proposition 5.22. That kind of discriminatory quota left the core nonempty, since it left the preferences substitutable, in the sense defined in Section 6.1.

6.3   Guide to the literature

This chapter takes its principle motivation from the paper of Kelso and Crawford (1982), which expanded on a paper by Crawford and Knoer (1981). Crawford and Knoer considered a discrete version (as well as the continuous version) of the assignment model (see Chapter 8), and developed a version of the deferred acceptance algorithm to prove the nonemptiness of the core. The earlier proof of Shapley and Shubik (1972) used a linear programming argument. (A polynomial time modification of Crawford and Knoer's algorithm is presented by Jones (1983).)

Kelso and Crawford formulated the model presented here in Section 6.2, as well as a continuous version of this model. They considered the deferred acceptance procedure with firms proposing, and remarked that because of the asymmetric nature of their model, the case with workers proposing seemed to pose greater difficulties. In Roth (1984c, 1985c) the discrete model was reformulated as the middle of three increasingly general models of one-to-one, many-to-one, and many-to-many matching, respectively. In the model of many-to-many matching, workers can take multiple jobs, just as firms can employ multiple workers, and all agents have preferences that obey the gross substitutes condition. The symmetry of this model allowed the existence of a worker-optimal pairwise stable outcome to be proved in a natural way for all three models.

Blair (1988) subsequently studied the symmetric model of many-to-many matching. He observed that in this model the set of (pairwise) stable outcomes no longer coincides with the core. Example 6.9 comes from his paper. An incorrect conclusion on this point was reached in Roth (1984c, 1985c), and readers of those papers are advised that they need to substitute "pairwise stable" for "stable" in the results concerning many-to-many matching. Blair showed that the (pairwise) stable outcomes preserve the lattice structure observed in all the other models presented here, when the partial order is chosen appropriately. However he observed that the lattices that arise in this model need no longer be distributive lattices.

As Examples 6.9 and 5.24 show, the identity between pairwise stable outcomes and core outcomes seems to be lost at the point at which we move from many-to-one to many-to-many matching. At this point, little more than that can be said: The structure of the core in the symmetric case, the conditions under which it is nonempty, and its relationship to the set of pairwise stable outcomes are all open questions. Of course, both the nonemptiness of the set of stable outcomes and its relationship to the core are connected to the assumption that preferences obey the gross substitutes condition. Kelso and Crawford's paper gives a careful interpretation of this assumption, which, as we have noted, is not a weak one. (That budget constraints could make the core empty in such models was noted by Mongell and Roth (1986), from which Example 6.12 is taken. That paper also noted that even when the core is nonempty, the presence of budget constraints can deprive it of much of the structure that the core has in the unconstrained case.) In general, this class of models has been less completely studied than most of the others in this volume, and so its limits are less well understood. However Crawford (1988) has generalized the comparative statics results of Kelso and Crawford (1982) concerning the addition of an agent on one side of the market, to markets of the sort considered in Roth (1984c, 1985c), using the generalization of the deferred acceptance algorithm introduced in Roth (1984c). Crawford's paper also contains a very clear argument concerning why two-sided matching models are particularly useful for modeling labor markets.

Sasaki and Toda (1986) consider a model of one-to-one matching in which there are externalities – that is, in which agents have preferences over matchings that may not depend only on whom they are matched to – and observe that pairwise stable outcomes need not be in the core in such a model, either. Prasad (1987) observes that the presence of externalities may substantially increase the computational complexity of such problems.

Another direction in which this kind of work might be extended is suggested by Kelso and Crawford. They note that their results can also be interpreted in terms of a one-sided market, containing no firms but only workers, who can form coalitions to produce output according to a technology that obeys the assumptions made about firms' technology in this model. Interpreted in this way, the assumptions of this model give conditions sufficient for the core of this one-sided game to be nonempty. In general, little is known about directions in which the two-sidedness of all the models we present here can be relaxed, and which of the results might be preserved. Some other directions in which two-sidedness can be relaxed while preserving nonemptiness of the core are explored by Kaneko and Wooders (1982) and Quinzii (1984).

The only empirical work we know of in which the kind of complex preferences considered in this chapter are an issue is that of Roth (1989b). There a variant of the simple model with substitutable preferences with which we begin this chapter, in Section 6.1, is suggested as a model of the kind of preferences that motivated Proposition 6.22, concerning quotas for different kinds of workers.

PART III

Models of one-to-one matching with money as a continuous variable

Chapters 7, 8, and 9 present three models of one-to-one matching in which money plays a prominent role. Unlike the model explored in Section 6.2, money will be modeled as a continuous rather than as a discrete variable, and this will let us employ a different set of mathematical tools. An even more important difference between the models of these chapters and those we have dealt with until now is that in these models, we are going to assume that the preferences individuals have for different matchings are basically monetary in nature.

Chapter 7 presents a model of exchange between one seller, with a single object to sell, and many buyers. This model will be simple enough so that the analysis can be conducted without much technical apparatus. We will explore it in enough detail to see that with one or two significant exceptions, the phenomena uncovered in our presentation of the marriage problem will also arise in these models with money. We will also relate this discussion to our discussion in Section 1.2 of strategic behavior in auctions, and particularly to the options available to coalitions of bidders.

Chapter 8 deals with a generalization to the case of many agents on each side of the market. We will see there that the existence of optimal stable (core) outcomes, and the lattice structure of the core, are related in this model to the basic theorems of linear programming. Chapter 9 deals with a further generalization that allows agents to have more complex tradeoffs in their preferences concerning whom they are matched with and how much they are paid. We will see that this model also permits us to prove parallels to most of the theorems derived for the marriage market. Our plan of presentation is to present much of the intuitive explanation when we discuss the simplest model in Chapter 7, while reserving the proofs of some of the mathematical results for the most general of these models in Chapter 9.

Thus the major theme of these chapters is that in two-sided models of one-to-one matching in which money plays a prominent role, the results very closely resemble the results obtained for the marriage problem. However these results are obtained in a different setting, using different mathematical tools. Just why the results should be so similar remains one of the open mathematical questions in this area.

CHAPTER 7

A simple model of one seller and many buyers

Consider a market consisting of a single seller, who owns one unit of an indivisible good, and n buyers, each of whom is interested in purchasing the object if the price is right. Each buyer b places a monetary value $\$r_b$ on the object, which is the maximum amount he or she is willing to pay, and the seller similarly places a value $\$r_s$ on the object, which is the price below which he or she will not sell. We call these monetary values the *reservation prices* of the agents. Each buyer has cash on hand sufficient to pay his or her reservation price.

We may interpret the reservation price as follows, for example. The seller has in hand an offer of $\$r_s$ from some outside party (not one of the buyers in the present market), who will buy the object at that price if it is offered by the seller. Each buyer (whom we may think of as a broker, rather than as a final consumer of the object) has in hand an offer of $\$r_b$ from a client who will purchase the object from the buyer at that price, should he or she obtain it in the market. Since the seller knows that he or she can earn at least $\$r_s$, the seller will not sell at a lower price. And since each buyer b knows that he or she can earn $\$r_b$ (but no more) by reselling the object, the buyer will not buy at a higher price.

So if the seller sells the object to buyer b at a price p (and if no other monetary transfers are made), the seller earns $\$p$, buyer b earns $\$(r_b - p)$, and all other buyers earn zero.

Note that we specify "if no other monetary transfers are made." This is a necessary caveat, since the agents in this model have money, and we do not rule out a priori any transfers of money between agents. That is, the model allows not only the conventional transfer of the purchase price from the successful buyer to the seller, but also allows side payments among buyers, or between the seller and more than one buyer. One of the things we will see in this model that applies to all the monetary models we study here, is that at stable outcomes (which correspond to the core) no such side payments are made. Thus the absence of these payments is not an assumption of the model, but emerges from the theory. However, when we consider the strategic options available to coalitions of buyers, we will see that such side payments reemerge.

7.1   The core of the game

Since we assume the preferences of the players for possible transactions can be translated into monetary benefits, we can represent this game in terms of the earnings of the players. To do so, define the set of players to be the set $N = \{1, ..., n, n+1\}$, where the buyers b are the players in the set $\{1, ..., n\}$, and the seller is player $n+1$. Let r_i for $i = 1, ..., n, n+1$ denote the reservation prices of the n buyers and one seller, respectively. For each nonempty subset S of N, called a *coalition of players*, a real-valued function v gives the worth of the coalition, $v(S)$, which is interpreted as the maximum amount of money the coalition can earn relying on the resources of coalition members. For this game, this "coalitional function" v can be defined as follows:

$v(S) = 0$ if S does not contain the seller, $n+1$;

$v(S) = \max\{r_i$ such that S contains $i\}$ if S does include the seller, $n+1$.

That is, no coalition that does not include the seller can earn any money in this market, since no profitable transactions can be completed without ownership of the object that is for sale. Any coalition that does include the seller can earn the maximum that any coalition member could earn if he or she could take possession of the object. So the maximum amount of money that can be earned in this game is $v(N)$, which equals the maximum of the reservation prices of the buyers and the seller. Since money is freely transferable in this model, knowing the worth of a coalition summarizes all the ways this wealth can be divided among the coalition members. A bit of notation will be helpful in what follows: for any vector of real numbers $x = (x_1, ..., x_n, x_{n+1})$, and any coalition of players S, denote the sum of the payoffs to the players in S by

$$x(S) \equiv \sum_{i \in S} x_i.$$

We can now define the set of all feasible monetary payoffs to the players in this game, under the assumption that money is freely transferable (in any amount) between players, to be the set X defined as follows:

$$X \equiv \{x = (x_1, ..., x_n, x_{n+1}) \in \mathbf{R}^{n+1} \text{ such that } x(N) \leq v(N)\}.$$

That is, a feasible payoff vector x is any one in which the sum of the payoffs to the players doesn't exceed the maximum feasible earnings of all the players. Note how this incorporates the assumption that money is freely transferable. If the only transfers that could occur were between the seller and the (one) successful buyer, the resulting vector x would have only two nonzero components. However this need not be the case if, say, the successful buyer makes payments to other buyers.

In the game theoretic literature a game given by the pair (N, v) denoting the set of players and a coalitional function (which is sometimes called a "characteristic function") is said to be modeled in *coalitional function form* with side payments. It is customary in the literature concerned with such games to concentrate on the subset $\bar{X}$ of X that is individually rational and Pareto optimal, which is called the set of *imputations*. That is,

$$\bar{X} \equiv \{x \in X \text{ such that } x_i \geq 0 \text{ for } i = 1, \ldots, n, \ x_{n+1} \geq r_{n+1},$$
$$\text{and } x(N) = v(N)\}.$$

We will see that when we are concerned with the core, it will not matter whether we consider the full set of feasible payoff vectors or only the set of imputations.

To define the core, we must define what it means for a payoff vector x to dominate another payoff vector y. Following Definition 3.1 we say that for any two feasible payoff vectors x and y, x *dominates* y if and only if there exists a coalition S of players such that

$$x_i > y_i \quad \text{for all } i \text{ in } S; \quad \text{and}$$

$$x(S) \leq v(S).$$

The first condition says that every member of the coalition prefers x to y (since in this model i prefers x to y if and only if $x_i > y_i$), and the second says that the coalition S has the ability to enforce (its part of) x, since it can afford to give to its members what they get at x. The *core* is of course the set of undominated payoff vectors, that is, it is the set $C \equiv$ $\{x \text{ in } X \text{ such that no } y \text{ in } X \text{ dominates } x\}$.

The following proposition, whose proof we leave to the reader, states that (for any coalitional function game with side payments) the core is a convex polyhedron (possibly empty) determined by $2^{n+1} - 1$ linear inequalities (one for each nonempty subset of the set of $n+1$ players).

Proposition 7.1. $C = \{x \text{ in } X \text{ such that } x(S) \geq v(S) \text{ for all coalitions } S\}.$

Before stating our main result for this section, we also define a payoff vector $x = (x_1, \ldots, x_n, x_{n+1})$ to be (pairwise) *stable* if it is individually rational and if there does not exist a buyer $i \in \{1, \ldots, n\}$ and a price p (a real number) such that $p > x_{n+1}$ and $r_i - p > x_i$. (If such a buyer and price did exist, the payoff vector x would be said to be blocked by the buyer and seller at that price, since they could make a transaction that would leave them both better off than at x.)

Within the framework of the general model, any example of this market is given by the vector of reservation prices $\mathbf{r} = (r_1, \ldots, r_n, r_{n+1})$. In order to characterize the core and the set of stable payoff vectors, it will be convenient to define a *reordering* of the players, $1^*, 2^*, \ldots, n+1^*$, so that $r_{1^*} \geq r_{2^*} \geq \cdots \geq r_{n+1^*}$. That is, under this alternative ordering, player 1^* is that player in N who has the highest reservation price (or one of the highest, if there is a tie), and $n+1^*$ is the player with the lowest reservation price.

We now state the main result of this section.

Theorem 7.2.

1. *For any reservation prices r, the core is nonempty.*

2. *If the seller does not have the highest reservation price (i.e., if $1^* \neq n+1$), and if $r_{1^*} > r_{2^*}$, then the core is given by $C = \{x \text{ in } X$ such that $x_{n+1} = p$ for $r_{2^*} \leq p \leq r_{1^*}$, $x_{1^*} = r_{1^*} - p$, and $x_i = 0$ for all i different from 1^* and $n+1\}$.*

   In any other case (i.e., if $1^* = n+1$ or $r_{1^*} = r_{2^*}$) then $C = \{(0, \ldots, 0, v(N))\}$.

3. *The set of stable payoff vectors equals the core.*

Note that the vector $x = (0, \ldots, 0, v(N))$, which gives all the benefits to the seller, is always in the core. (This is immediate either from the definition of domination or from Proposition 7.1, and proves statement 1 of the theorem.)

Statement 2 of the Theorem says that if some buyer has the unique highest reservation price (including the seller's), then the core divides the benefit between the seller and this buyer, who pays a price p between his or her own reservation price, r_{1^*}, and the next highest reservation price. No monetary transfers other than the payment of p from this buyer to the seller (in exchange for the transfer of the object from the seller to this buyer) are made. If the seller has the highest reservation price, or if the top two reservation prices are equal (whether they belong to two buyers or a buyer and a seller), then the only payoff vector in the core gives $v(N) = r_{1^*}$ to the seller, and zero to each buyer. (When the seller has the highest reservation price, this corresponds to the seller keeping the object; when more than one buyer has the same highest reservation price, this payoff vector corresponds to the seller selling the object to any of these buyers, at the full reservation price.) We leave to the reader to check that this fully describes the core, using either the definition of domination or the inequalities of Proposition 7.1.

Statement 3 of the theorem is now immediate. Since the price p obtained by the seller at a core payoff is always greater than or equal to $r_{2^\ast}$, there is no buyer who can profitably offer the seller a better price. And conversely, if a payoff vector x is not in the core, then it is either individually irrational or it is dominated via the coalition $\{1^\ast, n+1\}$, and so it is unstable.

We turn now to consider the relationship of these results to those obtained for the marriage model in Chaper 2.

7.1.1 Relationship to the results for the marriage model

That the set of stable outcomes is nonempty and equals the core obviously gives the same conclusion for this model as Theorems 2.8 and 3.3 do for the marriage market. We also have optimal stable outcomes for each side of the market, as in Theorem 2.12: In this simple model we can explicitly identify them, since the seller-optimal stable outcome corresponds to the selling price of $p = r_{1^\ast}$, and the buyer-optimal stable outcome corresponds to $p = r_{2^\ast}$. The preferences of the two sides of the market are opposed in the core (Theorem 2.13), since here the seller prefers p to be high and the buyers are all at least as well off when p is low. Note that where Theorem 2.13 requires an assumption of strict preferences, none was needed in Theorem 7.2. The effect of strict preferences in this model (and the more general models to follow) may arise out of the fact that we consider outcomes to be defined in the payoff space, where agents automatically have strict preferences (i.e., i is indifferent between x and y if and only if $x_i = y_i$).

In a similar vein, note that the core in this model is a lattice (as in Theorem 2.16), although a particularly simple one, since the outcomes are linearly ordered by p. So for this model we have no parallel to Theorem 3.9, which stated that all distributive lattices could arise as cores of the marriage markets. But in the models with many sellers considered in the next chapters, we will see a larger class of lattices.

The parallel to Theorem 2.22, which stated that in a marriage market the set of single people is the same for all stable matchings, is a little less exact. When $r_{1^\ast} > r_{2^\ast}$ there is a unique matching corresponding to each core outcome in this model, so the same result can be said to hold (i.e., only agent $1^\ast$ takes possession of the object at any core outcome). When $r_{1^\ast} = r_{2^\ast}$ then either $1^\ast$ or $2^\ast$ may take possession in the core, but the difference does not change the payoffs. If i receives payoff $x_i = 0$ at any core outcome, then he or she receives 0 at every core outcome. Note that it is essentially for this reason that we are able to suppress the matching in our description of this game, and concentrate on the set of payoff vectors.

Theorem 2.24 concerns the effect in the marriage market of having the men, say, extend their preferences to include new acceptable spouses. The parallel here would be to have some buyers raise their reservation prices (or of having the seller lower his or hers) so as to make some previously unacceptable transaction acceptable. As in the marriage problem, the effect of this can only be to improve the core from the point of view of the other side of the market.

Theorems 2.25 and 2.26 deal with the entry of new women, say, to the marriage market. Since the model examined in this chapter has only one seller, we will only consider for the moment the entry of new buyers. It is clear from Theorem 7.2 that the addition of a new reservation price r_{i^*} cannot hurt the seller (just as in Theorem 2.25) and if the new buyer turns out to have the highest reservation price, then the seller will be unambiguously helped (just as in Theorem 2.26).

Finally, the conclusion of Theorem 2.27, that the optimal stable outcome for either side of the market in the marriage model is weakly Pareto optimal for that side, most emphatically does not hold for the buyers in this market. This is because weak and strong Pareto optimality coincide in a model with side payments, and we saw in Example 2.31 that even in the marriage model, the optimal stable matchings need not be strongly Pareto optimal for either side. Here, all buyers could be made strictly better off than at the buyer-optimal stable outcome ($p = r_{2^*}$) if the successful buyer (1^*) got a price p' lower than r_{2^*}, and paid each of the other buyers a portion of the difference $r_{2^*} - p'$. We will see that this has important implications about the strategic opportunities facing coalitions.

7.2   Strategic questions

Our first result is an impossibility result, parallel to Theorem 4.4 for the marriage model. For our purposes in this section, a *stable matching mechanism* is a function h that, for every vector $\mathbf{r} = (r_1, \ldots, r_n, r_{n+1})$ of reservation prices, selects a payoff vector $h(\mathbf{r})$ in the core of the corresponding game $(N, v_\mathbf{r})$, where $v_\mathbf{r}$ is of course the characteristic function determined by the reservation prices $\mathbf{r}$.

Theorem 7.3. *No stable matching mechanism exists for which stating the true reservation price is a dominant strategy for every agent.*

The proof is simple: Suppose the reservation prices are such that $r_{1^*} > r_{2^*}$, and $1^* \neq n + 1$, that is, some buyer has the unique highest reservation price. Then there is a continuum of distinct outcomes in the core, corresponding to prices p in the interval $[r_{2^*}, r_{1^*}]$. Let $h(\mathbf{r})$ be any of these other than the buyer-optimal outcome; that is, let $h(\mathbf{r})$ correspond to a price $p^* > r_{2^*}$. Then buyer 1^* could have done better by stating a different reservation price, r_{1^*}' such that $r_{2^*} < r_{1^*}' < p^*$. In the corresponding game $v_{\mathbf{r}'}$ which differs from $v_{\mathbf{r}}$ only in the reservation price of 1^*, every core outcome (and therefore the outcome $h(\mathbf{r}')$) gives 1^* the good at a price less than p^*. (1^*'s profits, of course, continue to be measured from his or her true reservation price.) Similarly, if $h(\mathbf{r})$ in the original game is any outcome other than the seller-optimal outcome, that is, if it corresponds to a price $p^* < r_{1^*}$, then the seller can profitably misstate his or her reservation price to be r_{n+1}' in the interval $(p^*, r_{1^*}]$.

Three remarks about Theorem 7.3 and its proof are in order. First, impossibility theorems are strongest when stated on a limited class of games such as this: If no mechanism can satisfy the demands of the theorem on every game in this class, then, a fortiori, no mechanism can satisfy those demands on larger classes of games. So the conclusions of this theorem apply immediately to the more general models we consider in the next chapters. Second, by the revelation principle (recall Section 4.5.1), the theorem essentially implies that there is no way to formulate a strategic game of any kind (not merely one in which agents state their reservation prices) in which all the players have dominant strategies that always result in a core outcome. (If we enlarge the domain of games to include some with incomplete information about others' preferences, as in Section 4.5, then the implication is precise, i.e., we can delete the word "essentially" in the previous sentence. Otherwise we need to rule out, e.g., constant mechanisms, which pick some particular stable outcome no matter what the agents may do.) Finally, note that the proof of the theorem says that the possibility for some agent to profitably misrepresent his or her reservation price exists whenever the core contains more than one point, which is the same as the conclusion of Theorem 4.6 for the marriage model.

As in the marriage model, however, it is possible to make it a dominant strategy for either side of the market to state true reservation prices, by using the mechanism that always selects the optimal core outcome for that side. We will concentrate here on the mechanism that, for any stated reservation prices, chooses the buyer-optimal core outcome. This is a famous mechanism, variants of which are used in the auction of some U.S. government securities, for example. It is called the *sealed-bid, second-price auction* mechanism, and can be thought of as follows: Each buyer writes down a number (his or her bid, or stated reservation price) in an envelope, without knowing what number will be written down by any other buyer. The seller also writes down a number. All the envelopes are opened, and placed in order $r_1 \geq \cdots \geq r_{n+1^*}$, with the seller being player 1^*

only if his or her number is strictly greater than all the others, in which case there is no sale. Otherwise, buyer 1* receives the object, and pays the seller the price $p = r_{2^*}$.

This mechanism is sometimes also called a *Vickrey auction,* after the economist who first observed the following result in a famous 1961 paper.

Theorem 7.4 (Vickrey). *In a second-price, sealed-bid auction (which always yields the buyer-optimal core outcome in terms of the stated reservation prices), it is a dominant strategy for every buyer to state his or her true reservation price.*

It is easy to see why this is so. Consider a buyer b who states his true reservation price r_b, resulting in a vector **r** of stated reservation prices. Then, given the stated reservation prices of the others, b could not have helped himself, and could have hurt himself, if he had instead stated some reservation price different from the true one. If $b = 1^*$ with respect to these stated prices, that is, if his true reservation price is the highest stated price, then he gets the object at price $p = r_{2^*}$, which gives him a positive profit whenever r_{2^*} is strictly less than $r_b = r_{1^*}$. If he had stated a different reservation price, the outcome would not change at all so long as his stated price remains above r_{2^*}. But if he states a reservation price $r'_b < r_{2^*}$ (where by 2* we still mean the player with the second highest of the original reservation prices), buyer b will forego his profit, and receive a payoff of 0. (What happens when $r_b = r_{2^*}$ depends on what tie-breaking rule is used, but does not change the argument.) Now suppose that $b \neq 1^*$. Then b receives a payoff of 0, and would continue to do so for any stated preference $r'_b \leq r_{1^*}$. The only way b can change his payoff is (when $r_{1^*} > r_b$) by stating a reservation price $r'_b > r_{1^*}$, but in this case he buys the object at a price greater than his true reservation price, which gives him a negative profit. So it is a dominant strategy for each buyer to state his true reservation price.

Note that an important feature of this mechanism is that the price stated by a bidder determines if he or she is the winner, but does not determine the price he or she pays (as it would in a conventional first-price, sealed-bid auction in which submitting a bid equal to your reservation price is a domina*ted* strategy). Of course, this isn't the whole argument. A useful exercise for the reader to check that he or she has understood is to consider why a *third-price,* sealed-bid auction, that is, one at which buyer 1* receives the object but pays price r_{3^*}, does not make it a dominant strategy for each buyer to state his or her true reservation price.

As we noted in the proof of Theorem 7.3, it is not a dominant strategy for the *seller* to state his or her true reservation price when this mechanism is employed. In fact, stating the true reservation price is a best response for the seller only when all buyers state lower reservation prices. Otherwise, the best response for the seller is always to state a reservation price equal to r_1·. Thus there is a unique Nash equilibrium when the buyer-optimal stable mechanism is used, whose payoff is the seller-optimal outcome. This observation replaces the results of Theorems 4.6–4.17 and Theorem 4.11 for the marriage model.

Note that in this model, as in the marriage model, an equilibrium strategy for the side of the market without a dominant strategy may depend on a great deal of information that would in most practical circumstances be difficult to collect. (If the auctioneer knew the highest reservation price of the buyers, it wouldn't be necessary to conduct an auction.) So, as in the marriage model, when we deal with this kind of result we are running up against the limitations of the complete information model, which serves so well to illuminate other aspects of these models. To make progress on these kinds of equilibrium questions, models of incomplete information need to be considered (recall Section 4.5). In the case of auction markets, a good deal of progress has already been made in this way (see Section 7.3 for some references to the literature).

One of the reasons why the second-price, sealed-bid auction is of great interest is because of the relationship it has to more commonly observed *ascending bid* (also called "English") auctions. Consider a simple version of such an auction. The auctioneer keeps raising the price so long as two or more bidders indicate that they are still interested, and stops as soon as the next-to-last bidder drops out of the bidding. At that point, if the price is higher than the auctioneer's reservation price, the sale is made to the remaining bidder at the price at which the next-to-last remaining bidder dropped out. If the price at which the next-to-last bidder drops out is lower than the auctioneer's reservation price, the auctioneer acts as if there were a bidder who continued bidding until the auctioneer's reservation price is reached. Suppose for simplicity that the bidders cannot see which other bidders are still bidding. Then the problem facing a bidder b in this auction is to decide at what price to drop out of the auction. That is, he or she must decide on a single number, r_b, which together with the decisions of the other bidders (and the seller) determines the outcome in the same way as in the second-price, sealed-bid auction. So these two auctions are *strategically equivalent,* and the incentives facing the players are the same. (In most ascending bid auctions the situation is a little different than described here, but the idea presented here is approximately correct.) When all agents behave straightforwardly, the outcome is the buyer-optimal core outcome, and individual buyers have no incentive to behave strategically, but the seller (or auctioneer) does.

Of course, the incentive of the auctioneer to misrepresent his or her reservation price (by continuing to raise the price even after the reservation price has been exceeded and only one bidder remains) is especially strong if the auctioneer does not have to make the reservation price public at the beginning of the auction. Recall our discussion in Section 1.2 of the controversy over secret reservation prices in New York City. To fully discuss this question we would have to consider a model of incomplete information, since the chief problem facing an auctioneer who wishes to misrepresent the reservation price profitably is to estimate the (unknown) reservation prices of the bidders. (Again, recall Section 4.5 and see the references in Section 7.3.)

However, recall from the discussion in Section 1.2 that the auction houses also viewed secret reservation prices as a tool for combating bidder rings. We turn now to the strategic possibilities facing coalitions of bidders.

7.2.1   Bidder rings at auctions

It is clear that a *coalition* of bidders may be able, by suppressing some bids, to lower the price at which the object is sold in a second-price, sealed-bid auction, or an ascending bid auction (or, for that matter, in virtually any kind of auction). We will concentrate here on the second-price, sealed-bid auction. Consider a vector **r** of reservation prices for which the seller's reservation price is strictly less than the second highest, so that the sale price, $p = r_{2^*}$, is greater than the seller's (auctioneer's) reservation price r_{n+1}. Suppose the seller has the $(k+1)$st highest reservation price, that is, the seller is player $k+1^*$. Then the coalition consisting of bidders 1^* through k^* can, by having bidders 2^* through k^* submit bids less than the seller's reservation price, arrange for buyer 1^* to obtain the object at price $p' = r_{k+1^*} = r_{n+1} < r_{2^*}$.

Of course, if this was the end of the matter, buyer 1^* would benefit, but the coconspirators, 2^* through k^*, would not. However there is money in this model, so 1^* can share the wealth with the other $k-1$ members of the coalition, for example by paying each of them $[r_{2^*} - r_{k+1^*}]/k$. In this way all k members of the coalition $\{1^*, \dots, k^*\}$ profit equally from understating their reservation prices. Thus it is possible for a coalition of bidders acting together (a bidder ring) to profit from understating their bids and sharing the benefits among themselves, even though it is not possible for a single bidder acting alone to do better than to state his or her true reservation price.

Note how this compares with the results for the marriage model. In both models, it is a dominant strategy for an individual agent to state his or her true preferences when the choice consists of what preferences to state to the stable mechanism that chooses the optimal stable outcome for his or her side. In both models, no coalition of these agents may, by misstating their preferences, arrange so that they all do better under such a mechanism than when they all state their true preferences, *unless they are able to make side payments within the coalition*. That is, the conclusion of Theorem 4.10, as we were careful to formulate it in Chapter 4, is true in this model as well: If some coalition of bidders misstates its reservation prices so that the vector of reservations prices is $\bar{r}$ instead of r, then there is no outcome *in the core with respect to* $\bar{r}$ that all members of the ring prefer to the result of truthful revelation. This is because no money other than the purchase price is transferred at core outcomes. But as we have just seen, a coalition can profit by understating its preferences and then making side payments among its members.

Given that the results of a successful manipulation by a ring of bidders must lie outside of the core, we might wonder whether there exists a way of organizing the ring so that none of the members need worry that other members will fail to act as they have agreed. The following ingenious scheme, due to Graham and Marshall, shows that there is indeed such a way to organize a ring. They state the scheme as if it is organized by an artificial additional player, whom they call the "ring-center." We quote from their 1987 paper:

Prior to the main auction the risk neutral "ring-center" makes a fixed payment ..., P, to each of the ring members. Each of the k members of the ring then submits a sealed "reported bid" to the ring-center who determines the highest and second highest of these reported bids. The member of the ring who submitted the highest bid is then selected by the ring-center as the sole bidder and advised to submit this highest reported bid at the main auction. Ring members other than the sole bidder are advised to submit no bid or a zero bid at the main auction. Should the sole bidder win the item at the main auction he would pay the auctioneer the second highest of all bids submitted at the main auction. He would additionally pay the ring-center the difference between the second highest reported bid from the ring and the second highest of all bids at the main auction provided that this difference is positive.

In fact, Graham and Marshall state this scheme for an incomplete information environment. But the results that we are interested in involve dominant strategies, so we can continue to work with the complete information model, adapting the argument as needed, and keeping in mind the results about games of incomplete information about others' preferences from Proposition 4.25.

The observations we wish to make are these: If the choice facing each ring member consists only of what bids to submit to the ring and in the main auction, it is a dominant strategy for a ring member to state his or her true preferences to the ring, and to bid according to the coalition's plan at the main auction. Also, each member of the coalition is better off as a result of being a member of the ring than if he or she were to participate in the auction individually.

To see this, observe that the entire auction mechanism (including the coalition scheme) is still a second-price auction (although one in which some bidders are asked to submit bids on two occasions), since the person who submits the highest bid (either at the ring meeting or at the main auction) will have to pay the second highest bid submitted at either place. The argument used for Theorem 7.4 now suffices to show that no ring member can do better than to state his or her true preferences to the ring, nor to refrain from bidding if he or she is not the highest bidder in the ring. To see that a ring member is better off in the ring than out of it, observe that the ring member gets the object in either case only if his or her reservation price is the highest, but a ring member gets the additional payment P.

So far we have not specified P. So long as kP is not larger than the difference between the second highest reservation price within the ring and the highest reservation price outside of it, when this difference is positive the ring is self-financing. That is, the "ring-center" does not lose any money, since the winning bidder pays the ring-center (when no bidder plays a dominated strategy). (When this difference is negative, P must be 0.) So P will in general depend on the membership of the ring. (In Graham and Marshall's incomplete information formulation, P is the expected value of this amount, so the risk neutral ring-center comes out exactly even.) The important thing about the payment P is that it is made *in advance* to coalition members, before they have either stated their reservation prices to the coalition, or bid at the main auction. Consequently neither of these decisions can affect P, so the fact that they receive P doesn't have any effect on their incentives. (Note that if P were computed from the stated reservation prices, it might no longer be a dominant strategy to state true prices to the coalition.)

We remark in closing that this scheme is not meant to represent faithfully observed ring behavior. As Graham and Marshall emphasize, the scheme described here differs from observed behavior in several important respects. However this scheme for organizing a bidder ring gives some idea of the rich strategic possibilities open to coalitions of bidders. These strategic possibilities arise even when a bidder-optimal stable mechanism is employed, in sharp contrast to the situation facing individual bidders in that case.

7.3   Guide to the literature

The results presented in Section 7.1 are largely a straightforward application of simple game-theoretic principles laid out by von Neumann and Morgenstern (1944), who defined games in coalitional (characteristic) function form with side payments, as well as the notion of one payoff vector dominating another.

The results in Section 7.2 draw on more modern sources. Vickrey's (1961) paper is often credited with being the first result of the kind, and many other dominant strategy mechanisms of one sort or another have been considered since. We have drawn heavily on the recent papers of Graham and Marshall (1984, 1987); some further work is found in Graham, Marshall, and Richard (1987a, b). These papers develop models of auction behavior under the natural assumption of incomplete information. Some of the basic work on noncooperative behavior in auctions modeled as games of incomplete information is found in Milgrom and Weber (1982) and Myerson (1981, 1983). A number of experimental investigations concerning auction behavior have been conducted; see Roth (1988a) for references. The general theory of games of incomplete information opens up a number of avenues of exploration not considered here. Along these lines, useful ways to think about auctions under conditions of incomplete information are presented by Holmstrom and Myerson (1983) and Cremer and McLean (1985). Ashenfelter (1989) describes some further observations of auction practices, and suggests a number of considerations that have often been left out of theoretical models in the literature.

We noted in the complete information model considered here that the equilibrium of the buyer-optimal stable mechanism is the seller-optimal outcome. Thomson (1986) notes that this result has parallels in other kinds of models than those considered here.

CHAPTER 8

The assignment game

8.1   The formal model

This chapter presents a model in which there may be many sellers and many buyers, or many firms and workers. Formally, there are two finite disjoint sets of players P and Q, containing m and n players, respectively. Members of P will sometimes be called P-agents and members of Q called Q-agents, and the letters i and j will be reserved for P- and Q-agents, respectively. Associated with each possible partnership (i, j) in $P \times Q$ is a nonnegative real number α_{ij}. A game in coalitional function form with side payments is determined by (P, Q, α), with the numbers α_{ij} being equal to the worth of the coalitions $\{i, j\}$ consisting of one P-agent and one Q-agent. The worth of large coalitions is determined entirely by the worth of the pairwise combinations that the coalition members can form. That is, the coalitional function v is given by

$v(S) = \alpha_{ij}$ if $S = \{i, j\}$ for i in P and j in Q;

$v(S) = 0$ if S contains only P-agents or only Q-agents; and

$v(S) = \max(v(i_1, j_1) + v(i_2, j_2) + \cdots + v(i_k, j_k))$ for arbitrary coalitions S, with the maximum to be taken over all sets $\{(i_1, j_1), \ldots, (i_k, j_k)\}$ of k distinct pairs in $S_P \times S_Q$, where S_P and S_Q denote the sets of P- and Q-agents in S (i.e., the intersection of the coalition S with P and with Q) respectively. Of course the number k of pairs in this maximization problem cannot exceed the minimum of $|S_P|$ and $|S_Q|$.

So the rules of the game are that any pair of agents (i, j) in $P \times Q$ can together obtain α_{ij}, and any larger coalition is valuable only insofar as it can organize itself into such pairs. The members of any coalition may divide among themselves their collective worth in any way they like. An imputation of this game is thus a nonnegative vector (u, v) in $R^m \times R^n$ such that $\sum_{i \in P} u_i + \sum_{j \in Q} v_j = v(P \cup Q)$. The easiest way to interpret this is to take the quantities α_{ij} to be amounts of money, and to assume that agents' preferences are concerned only with their monetary payoffs.

We might think of this kind of game as arising from the multiseller generalization of the model of Chapter 7, where P is a set of potential buyers of some objects offered for sale by the set Q of potential sellers, and each seller owns and each buyer wants exactly one indivisible object. If each seller has a reservation price of zero, then the α_{ij}'s represent each buyer i's reservation price for the object offered by seller j. In this case if buyer i buys from seller j at a price p, and if no other monetary transfers are made or received by i and j, then the resulting utilities to the two agents are $u_i = \alpha_{ij} - p$ and $v_j = p$. More generally, if each seller j has a reservation price c_j, and each buyer i has a reservation price r_{ij} for object j, we may take α_{ij} to be the potential gains from trade between i and j; that is $\alpha_{ij} = \max\{0, r_{ij} - c_j\}$. In this case if buyer i buys object j from seller j at a price p, and if no other monetary transfers are made, the utilities are $u_i = r_{ij} - p$ and $v_j = p - c_j$. (It will be convenient to normalize each seller's utility function in this way, with the utility of keeping his own object being zero rather than c_j as in the previous chapter, so that these utilities u_i and v_j sum to α_{ij}. There is no loss of generality in doing so.) Note that transfers between agents are not restricted to those between buyers and sellers; for example, buyers may make transfers among themselves as in the bidder rings of Section 7.2.1.

Of course, in a similar way we can think of the P- and Q-agents as being firms and workers, and so on. As in the marriage model, we look here at the simple case of one-to-one matching, with firms constrained to hire at most one worker. In such a case, the α_{ij}'s represent some measure of the joint productivity of the firm and worker, and transfers between a matched firm and worker represent salary. Transfers can also take place between workers (as when workers form a labor union in which the dues of employed members help pay unemployment benefits to unemployed members) or between firms.

Note that since money is freely transferable and since each agent's preferences are assumed to be essentially monetary in nature, we are assuming that no agent has strict preferences. That is, for every pair of objects and any buyer, there is a pair of prices that makes the buyer indifferent between purchasing either of the objects.

The evaluation of the maximization problem to determine $v(S)$ for a given matrix α is called an *optimal assignment problem* or simply an *assignment problem,* so games of this form are called *assignment games.* We will be particularly interested in the value of the coalition $P \cup Q$, since $v(P \cup Q)$ equals the maximum total payoff available to the players in this game, and hence determines the Pareto set and the set of imputations.

Consider the following linear programming (LP) problem P_1:

Maximize $\sum\limits_{i,j} \alpha_{ij} \cdot x_{ij}$

subject to (a) $\sum\limits_i x_{ij} \leq 1$

(b) $\sum\limits_j x_{ij} \leq 1$

(c) $x_{ij} \geq 0.$

Note that constraints (a), (b), and (c) are almost the same as constraints (1), (2), and (4) in Theorem 3.2.1. (The difference is that the inequalities in (a) and (b) allow agents to be unmatched.) So we may interpret x_{ij} as, for example, the probability that a partnership (i, j) will form. Then the linear inequalities of type (a), one for each j in q, say that the probability that j will be matched to some i cannot exceed 1. The inequalities of form (b), one for each i in P, say the same about the probability that i will be matched.

It can be shown as in Section 3.2.4 (see, e.g., Dantzig 1963, 318) that there exists a solution of this LP problem that involves only values of zero and one. (The extreme points of systems of linear inequalities of the form (a), (b), and (c) have integer values of x_{ij}; i.e., each x_{ij} equals zero or one.) Thus the fractions artificially introduced in the LP formulation disappear in the solution and the (continuous) LP problem is equivalent to the (discrete) assignment problem for the coalition of all players, that is, the determination of $v(P \cup Q)$. Then $v(P \cup Q) = \sum \alpha_{ij} \cdot x_{ij}$, where x is an optimal solution of the LP problem.

Definition 8.1. *A feasible assignment for (P, Q, α) is a matrix $x = (x_{ij})$ (of zeros and ones) that satisfies* (a), (b), *and* (c) *above.*

Then using the interpretation of x given above we can say that $x_{ij} = 1$ if i and j form a partnership and $x_{ij} = 0$ otherwise. If $\sum_j x_{ij} = 0$, then i is *unassigned*, and if $\sum_i x_{ij} = 0$, then j is likewise unassigned. A feasible assignment x corresponds exactly to a matching μ as defined in Definition 2.1, with $\mu(i) = j$ if and only if $x_{ij} = 1$. And it is equivalent to say that an agent i or j is unassigned at x or is unmatched (single) at μ.

Any solution of the preceding LP problem is called an *optimal assignment*.

Definition 8.2. *A feasible assignment x is optimal for (P, Q, α) if, for all feasible assignments x', $\sum_{i,j} \alpha_{ij} \cdot x_{ij} \geq \sum_{i,j} \alpha_{ij} \cdot x'_{ij}$.*

An assignment problem always has a solution, since there are only a finite number of assignments. For example, consider the assignment problem given by

$$\alpha = \begin{pmatrix} 10 & 12 & 7 \\ 6 & 8 & 2 \\ 5 & 5 & 9 \end{pmatrix}.$$

There are two optimal assignments given by

$$x = \begin{pmatrix} 1 & 0 & 0 \\ 0 & 1 & 0 \\ 0 & 0 & 1 \end{pmatrix} \quad \text{and} \quad x' = \begin{pmatrix} 0 & 1 & 0 \\ 1 & 0 & 0 \\ 0 & 0 & 1 \end{pmatrix}$$

with value $\alpha_{11} + \alpha_{22} + \alpha_{33} = \alpha_{12} + \alpha_{21} + \alpha_{33} = 27$.

Definition 8.3. *The pair of vectors* (u, v), *with u in R^m and v in R^n, is called a **feasible payoff** for (P, Q, α) if there is a feasible assignment x such that*

$$\sum_{i \in P} u_i + \sum_{j \in Q} v_j = \sum_{\substack{i \in P \\ j \in Q}} \alpha_{ij} \cdot x_{ij}.$$

In this case we say (u, v) and x are *compatible* with each other, and we call $((u, v); x)$ a *feasible outcome*. Note again that a feasible payoff vector may involve monetary transfers between agents who are not assigned to one another.

As in the models of earlier chapters, the key notion is that of stability.

Definition 8.4. *A feasible outcome* $((u, v); x)$ *is **stable** (or the payoff* (u, v) *with an assignment x is stable) if*

(i) $u_i \geq 0, \quad v_j \geq 0$

(ii) $u_i + v_j \geq \alpha_{ij}$ *for all* (i, j) *in* $P \times Q$.

Condition (i) (individual rationality) reflects that a player always has the option of remaining unmatched (recall that $v(i) = v(j) = 0$ for all individual agents i and j). Condition (ii) requires that the outcome is not blocked by any pair: If (ii) is not satisfied for some agents i and j, then it would pay them to break up their present partnership(s) (either with one another or with other agents) and form a new partnership together, because this could give them each a higher payoff.

From the definition of feasibility and stability it follows that

Lemma 8.5.   *Let* $((u,v),x)$ *be a stable outcome for* (P,Q,α). *Then*

(i) $u_i + v_j = \alpha_{ij}$ *for all pairs* (i,j) *such that* $x_{ij} = 1$

(ii) $u_i = 0$ *for all unassigned* i, *and* $v_j = 0$ *for all unassigned* j *at* x.

Proof: Let R (respectively S) be the set of all unassigned i (respectively j) at x. Then by feasibility of $((u,v)x)$:

$$\sum_P u_i + \sum_Q v_j = \sum_{P \times Q} (u_i + v_j)x_{ij} + \sum_{i \in R} u_i + \sum_{i \in S} v_j = \sum_{P \times Q} \alpha_{ij} \cdot x_{ij}.$$

Now apply the definition of stability.

The lemma implies that at a stable outcome, the only monetary transfers that occur are between P- and Q-agents who are matched to each other. (Note that this is an implication of stability, not an assumption of the model.)

8.2   The core of the assignment game

Consider the LP problem P_1^* that is the dual of P_1, that is, the LP problem of finding a pair of vectors (u,v) in $R^m \times R^n$, that minimizes the sum

$$\sum_{i \in P} u_i + \sum_{i \in Q} v_j$$

subject, for all i in P and j in Q, to

(a*) $u_i \geq 0,\ v_j \geq 0$

(b*) $u_i + v_j \geq \alpha_{ij}$.

Because we know that P_1 has a solution, we know also that P_1^* must have an optimal solution. A fundamental duality theorem (see Dantzig, 1963, 129) asserts that the objective functions of these dual LP's must attain the same value. That is, if x is an optimal assignment and (u,v) is a solution of P_1^*, we have that

$$\sum_{i \in P} u_i + \sum_{i \in Q} v_j = \sum_{P \times Q} \alpha_{ij} \cdot x_{ij} = v(P \cup Q). \tag{8.1}$$

This means that $((u,v),x)$ is a feasible outcome. Moreover, $((u,v),x)$ is a stable outcome for (P,Q,α) since (a*) ensures individual rationality and $u_i + v_j \geq \alpha_{ij}$ for all (i,j) in $P \times Q$ by (b*).

On the other hand, condition (b*) says that

$$u_i + v_j \geq v(i,j)   \text{ for all } i \text{ in } P, j \text{ in } Q.$$

It follows, by the definition of $v(S)$, that for any coalition $S = S_P \cup S_Q$, where S_P is contained in P and S_Q in Q,

$$\sum_{i \in S_P} u_i + \sum_{j \in S_Q} v_j \geq v(S). \tag{8.2}$$

But (8.1) and (8.2) are exactly how the core of the game is determined (recall Proposition 7.1): (8.1) ensures the feasibility of (u, v) and (8.2) ensures its nonimprovability by any coalition. Conversely, any payoff vector in the core, that is satisfying (8.1) and (8.2), satisfies the conditions for a solution to P_1^*.

Hence we have shown that

Theorem 8.6 (Shapley and Shubik). *Let (P, Q, α) be an assignment game. Then*

(a) *the set of stable outcomes and the core of (P, Q, α) are the same.*
(b) *the core of (P, Q, α) is the (nonempty) set of solutions of the dual LP of the corresponding assignment problem.*

The following two corollaries make clear why, in contrast to the discrete models considered earlier, we can concentrate here on the payoffs to the agents rather than on the underlying assignment (matching).

Corollary 8.7. *If x is an optimal assignment, then it is compatible with any stable payoff (u, v).*

Proof: Immediate from the fact that if (u, v) is a stable payoff, then it satisfies (8.1) for any optimal assignment.

Corollary 8.8. *If $((u, v), x)$ is a stable outcome, then x is an optimal assignment.*

Proof: Immediate from the fact that

$$\sum_j u_i + \sum_j v_j = v(P \cup Q) = \sum_{i,j} \alpha_{ij} \cdot x_{ij}.$$

As in the marriage model, if i prefers a stable payoff (u, v) to another stable payoff (u', v'), his or her mate(s) will prefer (u', v') (recall Corollary 2.21).

Proposition 8.9. *Let $((u, v), x)$ and $((u', v'), x')$ be stable outcomes for (P, Q, α). Then if $x'_{ij} = 1$, $u'_i > u_i$ implies $v'_j < v_j$.*

Proof: Suppose $v'_j \geq v_j$. Then $\alpha_{ij} = u'_i + v'_j > u_i + v_j \geq \alpha_{ij}$, which is a contradiction.

Just as Proposition 8.9 shows how the interests of P- and Q-agents are opposed in the core, the following theorem shows that among themselves, the P-agents and Q-agents have common interest in the core. Specifically, as in the marriage market, the core is a lattice; that is, the greatest lower (or least upper) bound to any two points in the core is also in the core (recall Theorems 2.16 and 3.8).

Define the partial order $(u', v') >_P (u, v)$ if $u_i' > u_i$ for all i in P and $u_i' > u_i$ for at least one i in P. It follows from Proposition 8.9 that for stable outcomes, if $(u', v') >_P (u, v)$ then $v_j' \le v_j$ for all j in Q. Then we have

Theorem 8.10 (Shapley and Shubik). *The core of the assignment game endowed with the partial order $\ge_P$ forms a complete lattice (dual to the lattice with ordering $\ge_Q$).*

Proof: Let (u, v) and (u', v') be any two payoff vectors in the core. Let x be some optimal assignment. Let

$$\underline{u}_i = \min\{u_i, u_i'\} \qquad \underline{v}_j = \min\{v_j, v_j'\}$$
$$\bar{u}_i = \max\{u_i, u_i'\} \qquad \bar{v}_j = \max\{v_j, v_j'\}.$$

We will show that $((\underline{u}, \bar{v}), x)$ and $((\bar{u}, \underline{v}), x)$ are also in the core. For any i and j we have either

$$\underline{u}_i + \bar{v}_j = u_i' + \bar{v}_j \ge u_i' + v_j' \ge \alpha_{ij} \quad \text{or}$$
$$\underline{u}_i + \bar{v}_j = u_i + \bar{v}_j \ge u_i + v_j \ge \alpha_{ij}.$$

By Corollary 8.7, (u, v) and (u', v') are compatible with x. Clearly $\underline{u}_i \ge 0$ and $\bar{v}_j \ge 0$. It remains to show that $\sum_i \underline{u}_i + \sum_j \bar{v}_j = v(P \cup Q)$. But it is immediate, from Proposition 8.9 and Lemma 8.5, that if $x_{ij} = 1$ then

$$\underline{u}_i + \bar{v}_j = u_i' + v_j' = \alpha_{ij} \quad \text{or}$$
$$\underline{u}_i + \bar{v}_j = u_i + v_j = \alpha_{ij}.$$

Hence

$$\sum_i \underline{u}_i + \sum_j \bar{v}_j = \sum_{i,j} \alpha_{ij} \cdot x_{ij} = v(P \cup Q).$$

Analogously, $(\bar{u}, \underline{v})$ is stable. Hence we have shown that the core is a lattice. Since it is a convex polytope it is also a compact set, from which it follows that it is a complete lattice. $\quad$

As in the marriage market, this implies the existence of P- and Q-optimal stable outcomes. That is, there is a vertex in the core at which every player from one side gets the maximum payoff and every agent from the other side gets the minimum payoff. There is another vertex with symmetric properties. This is an immediate consequence of Theorem 8.10 and Proposition 8.9.

Theorem 8.11 (Shapley and Shubik). *There is a P-optimal stable payoff* $(\bar{u}, \underline{v})$, *with the property that for any stable payoff* (u, v), $\bar{u} \geq u$ *and* $\underline{v} \leq v$; *there is a Q-optimal stable payoff* $(\underline{u}, \bar{v})$ *with symmetrical properties.*

8.3   A multiobject auction mechanism

In this section we will interpret P as a set of *bidders* and Q as a set of *objects*. Each object j has a reservation price of c_j. The value of object j to bidder i is $\alpha_{ij} \geq 0$. A feasible price vector p is a function from Q to R^+ such that $p_j = p(j)$ is greater than or equal to c_j. As a notational convention we will also assume in this section that Q contains an artificial "null object," O, whose value α_{iO} is zero to all bidders and whose price is always zero. Then if a bidder is unmatched we will say that he or she is assigned to O. (More than one bidder may be assigned to O.) The *demand set* of a bidder i at prices p is defined by

$$D_i\{p\} = \{j \in Q; \ \alpha_{ij} - p_j = \max_{k \in Q}\{\alpha_{ik} - p_k\}\}.$$

The price vector p is called *quasi-competitive* if there is a matching μ from P to Q such that if $\mu(i) = j$ then j is in $D_i(p)$, and if i is unmatched under μ then O is in $D_i(p)$. Thus at quasi-competitive prices p each buyer can be assigned to an object in his or her demand set. The matching μ is said to be *compatible* with the price p. The pair (p, μ) is a *competitive equilibrium* if p is quasi-competitive, μ is compatible with p, and $p_j = c_j$ for all $j \notin \mu(P)$. Thus at a competitive equilibrium, not only does every buyer get an object in his or her demand set, but no unsold object has a price higher than its reservation price. If (p, μ) is a competitive equilibrium, p will be called a *competitive* or an *equilibrium* price vector.

It is easy to verify that if (p, μ) is a competitive equilibrium, then the corresponding payoffs (u, v) are stable (where $u_i = \alpha_{ij} - p_j$ and $v_j = p_j - c_j$ for $j = \mu(i)$). The existence of a P-optimal stable payoff is equivalent to the statement that there is a unique vector of equilibrium prices that is optimal for the P-agents, in the sense that it is at least as small in every component as any other equilibrium price vector. This price is called the *minimum equilibrium price*. We will describe an algorithm for computing this price, which is an auction mechanism that generalizes the Vickrey second-price auction described in Chapter 7. (Note that the Vickrey auction of a single object also produces the minimum equilibrium price.) As we will see in Section 8.4, one important property of the single-object auction that generalizes to the multiobject case is that submitting true valuations is a dominant strategy for the bidders.

To describe the mechanism, we will make use of the following well-known result from graph theory. Let B and C be two finite disjoint sets (e.g., of buyers and objects, respectively). For each i in B, let D_i be a subset of C (e.g., D_i is i's demand set at some set of prices). A *simple assignment* is an assignment of objects to buyers such that each buyer i is assigned exactly one object j such that j is in D_i, and each object is assigned to at most one buyer. (So a simple assignment assigns an object to every buyer but may not assign every object to a buyer.) Then it is apparent that if a simple assignment exists, each buyer in every subset B' of B must be matched to a different object, so there must be at least as many objects in $D(B') \equiv \bigcup_{i \in B'} D_i$ as there are buyers in B'. Hall's theorem says that this necessary condition is also sufficient.

Theorem 8.12 (Hall's theorem).   *A simple assignment exists if and only if, for every subset B' of B, the number of objects in $D(B')$ is at least as great as the number of buyers in B'.*

The auction mechanism for the multiobject case that we will now present produces the minimum price equilibrium in a finite number of steps.

We will take all prices and valuations to be integers. At the first step of the auction the auctioneer announces an initial price vector, $p(1)$, equal to the vector c of reservation prices. Each bidder "bids" by announcing which object or objects (including the null object O) are in his or her demand set at price $p(1)$.

Step $(t+1)$:   After the bids are announced, if it is possible to match each bidder to an object in his or her demand set at price $p(t)$ the algorithm stops. If no such matching exists, Hall's theorem implies that there is some *overdemanded* set, that is, a set of objects such that the number of bidders demanding only objects in this set is greater than the number of objects in the set. The auctioneer chooses a *minimal* overdemanded set (i.e., an overdemanded set S such that no strict subset of S is an overdemanded set) and raises the price of each object in the set by one unit. All other prices remain at the level $p(t)$. This defines $p(t+1)$. (Note that the nonexistence of the matching implies the minimal overdemanded set does not contain the null object O, since we allow any number of agents to be matched to O if O is in their demand sets.)

It is clear that the algorithm stops at some step t, because as soon as the price of an object becomes higher than any bidder's valuation for it, no bidder can demand it. It follows that the final price obtained by this algorithm is a quasi-competitive price vector. Indeed it is the minimum equilibrium price vector, although this fact is not so obvious.

Theorem 8.13 (Demange, Gale, Sotomayor). *Let p be the price vector obtained from the auction mechanism. Then p is the minimum quasi-competitive price.*

Proof: Suppose instead that there exists a quasi-competitive price q such that $p \nleq q$. Now at step $t = 1$ of the auction we have $p(1) = c$ so $p(1) \leq q$. Let t be the last step of the auction at which $p(t) \leq q$ and let $S_1 = \{j; p_j(t+1) > q_j\}$. Let S be the minimal overdemanded set whose prices are raised at stage $t + 1$, thus $S = \{j; p_j(t+j) > p_j(t)\}$, so S_1 is contained in S. Furthermore $q_j = p_j(t)$ for all j in S_1 (since we are working with all integers). We will show that $S - S_1$ is nonempty and overdemanded, hence S is not a minimal overdemanded set, contrary to the rules of the auction.

Define $T = \{i; D_i(p(t))$ is contained in $S\}$. That S is overdemanded means exactly that

$$|T| > |S|. \tag{1}$$

Define $T_1 = \{i \in T;$ the set of objects in S_1 demanded by i at price $p(t)$ is nonempty$\}$.

We claim that $D_i(q)$ is contained in S_1 for all i in T_1. Indeed, choose j in S_1 and in $D_i(p(t))$. If $k \notin S$, then i prefers j to k at price $p(t)$ because i is in T, but $p_k(t) \leq q_k$ and $p_j(t) = q_j$. So i prefers j to k at price q. On the other hand, if k is in $S - S_1$, then i likes j at least as well as k at price $p(t)$, but $p_k(t) < p_k(t+1) \leq q_k$ (and, again, $p_j(t) = q_j$) so i prefers j to k at price q, as claimed. Now since q is quasi-competitive there are no overdemanded sets at price q so

$$|T_1| \leq |S_1|. \tag{2}$$

Now from (1) and (2), $|T - T_1| > |S - S_1|$ so $T - T_1 \neq \emptyset$ and $T - T_1 = \{i \in T; D_i(p(t)) \in S - S_1\}$. So $S - S_1 \neq \emptyset$ and $S - S_1$ is overdemanded, giving the desired contradiction.

Theorem 8.14 (Demange, Gale, Sotomayor). *If p is the minimum quasi-competitive price, then there is a matching μ^* such that (p, μ^*) is an equilibrium (so p is a competitive price vector).*

Proof: Let μ be a matching corresponding to p. Call an object j *overpriced* if it is unmatched by μ but $p_j > c_j$. If (p, μ) is not an equilibrium, there is at least one overpriced object. We will give a procedure for altering μ so as to eliminate overpriced objects. For this purpose we construct a directed graph whose vertices are $P \cup Q$. There are two types of arcs. If $\mu(i) = j$ there is an arc from i to j. If j is in $D_i(p)$ there is an arc from j to i. Now let k be an overpriced object. Then k is in $D_i(p)$ for some i, for if not we could decrease p_k and still have quasi-competitive prices, which contradicts the minimality of p. Let $\bar{P} \cup \bar{Q}$ be all vertices that can be reached by a directed path starting from k.

Case 1: $\bar{P}$ contains an unmatched bidder, i. Let $(k, i_1, j_2, i_2, j_3, i_3, \ldots, j_\ell, i)$ be a path from k to i. Then we may change μ by matching i_1 to k, i_2 to $j_2, \ldots, i$ to j_ℓ. The matching is still competitive and k is no longer overpriced so the number of overpriced objects has been reduced.

Case 2:   All i in $\bar{P}$ are matched. Then we claim that there must be some j in $\bar{Q}$ such that $p_j = c_j$, for suppose not. By definition of $\bar{P} \cup \bar{Q}$ we know that if $i \notin \bar{P}$ then i does not demand any object in $\bar{Q}$. Therefore we can decrease the price of each object in $\bar{Q}$ by some positive δ and still have quasi-competitiveness, contradicting the minimality of p. So choose j in $\bar{Q}$ such that $p_j = c_j$ and let $(k, i_1, j_2, i_2, \ldots, j_\ell, i_\ell, j)$. Again change μ by matching i_1 to k, i_2 to $j_2, \ldots$, leaving j unmatched. Again the number of overpriced objects has been reduced.

8.4   Incentives

Denote by $(\bar{u}, \underline{v})$ the P-optimal stable payoff for the market $M = (P, Q, \alpha)$. In this section it will continue to be convenient to think of P-agents as buyers, and Q-agents as sellers. (But we will no longer speak of unmatched buyers as demanding an artificial null object O, nor will we continue to take all prices to be integers.) For simplicity we will take the reservation prices c to all be zero, so $\underline{v}$ is the minimum equilibrium price vector. Let v be the coalitional function of the game, that is, for every S contained in P and R contained in Q, $v(S, R) = \max \sum_{S \times R} \alpha_{ij} x_{ij}$, for all assignments x. The demand set of buyer i at prices $\underline{v}$ is defined by $D_i(\underline{v}) = \{j \in Q$ such that $\alpha_{ij} - \underline{v}_{ij} \geq 0$, $\alpha_{ij} - \underline{v}_{ij} = \max_{k \in Q}\{\alpha_{ik} - \underline{v}_{ik}\}\}$. (Note that now that we have dispensed with the null object, the demand set of a buyer may be empty.)

The following lemma shows a critical way in which the Vickrey second-price auction is generalized by the mechanism that sets prices equal to $\underline{v}$ (i.e., that gives buyers their optimal stable outcome). Both mechanisms give buyers their marginal contribution to coalitional values.

Lemma 8.15 (Demange; Leonard).   *For all i in P,*

$$\bar{u}_i = v(P, Q) - v(P - \{i\}, Q).$$

Proof: Let x be an optimal assignment for $M = (P, Q, \alpha)$. Construct a graph whose vertices are $P \cup Q$. There are two kinds of arcs. If $x_{ij} = 1$ there is an arc from i to j. If j is in $D_i(\underline{v})$ and $x_{ij} = 0$ there is an arc from j to i. Let j be an object whose price is greater than zero. Then there is an oriented path starting from j and ending at an unmatched buyer or at an object of price zero. To see this, suppose there is no such path, and denote by S and T the sets of objects and buyers, respectively, that can be reached from j. Then $\underline{v}_k > 0$ for all k in S. Furthermore, if $i \notin T$, then there is no object in S that is demanded by i at price $\underline{v}$. (If k is demanded by i then there is an arc from k to i if $x_{ik} = 0$, or an arc from i to k if $x_{ik} = 1$. In both cases, if i is not in T, k cannot be in S.) Then we can decrease $\underline{v}_k$ for all k in S, and still have an equilibrium, which contradicts the minimality of $\underline{v}$.

So, let i' be any buyer. If i' is assigned to some object j_1, we may consider a path c beginning at j_1 and ending at an unmatched buyer i_s or at an object k of price zero. (Note that k might be j_1.) That is, $c = (j_1, i_1, j_2, i_2, \ldots, j_s, i_s)$ or $c = (j_1, i_1, j_2, i_2, \ldots, j_s, i_s, k)$. Consider now the assignment x' in $M' = (P - \{i'\}, Q, \alpha)$ that assigns j_1 to i_1, j_2 to $i_2, \ldots, j_s$ to i_s, and that leaves k unmatched if k is in the path, and that otherwise agrees with x on every buyer in $P - \{i'\}$ who is not in the path. We claim the outcome $((u^*, \underline{v}); x')$ is stable for M', where $u_i^* = \bar{u}_i$ for all $i \neq i'$. This is immediate from the fact that $x'_{i_t, j_t} = 1$, j_t is demanded by i_t at price $\underline{v}_{j_t}$ for all $t = 1, \ldots, s$, and $((\bar{u}, \underline{v}), x)$ is stable for (P, Q, α). Then x' is an optimal assignment for M', so

$$\sum_{\substack{i \neq i' \\ j \in Q}} \alpha_{ij} x'_{ij} = v(P - \{i'\}, Q). \tag{a}$$

On the other hand,

$$\sum_{\substack{i \neq i' \\ j \in Q}} \alpha_{ij} x'_{ij} = \sum_i u_i^* + \sum_j \underline{v}_j = \sum_{i \neq i'} \bar{u}_i + \sum_j \underline{v} = v(P, Q) - \bar{u}_{i'}. \tag{b}$$

From (a) and (b) we obtain $\bar{u}_{i'} = v(P, Q) - v(P - \{i'\}, Q)$, which completes the proof.

Let x' be any optimal assignment for $(P - \{i\}, Q - \{j\}, \alpha)$, where i is assigned to j under the optimal assignment x for (P, Q, α). Then

$$\sum_{\substack{\ell \neq i \\ k \neq j}} \alpha_{\ell k} \cdot x'_{\ell k} + \alpha_{ij} \leq \sum_{\substack{\ell \neq i \\ k \neq j}} \alpha_{\ell k} \cdot x_{\ell k} + \alpha_{ij},$$

from optimality of x. Then

$$\sum_{\substack{\ell \neq i \\ k \neq j}} \alpha_{\ell k} x'_{\ell k} \leq \sum_{\substack{\ell \neq i \\ k \neq j}} \alpha_{\ell k} \cdot x_{\ell k}. \tag{1}$$

On the other hand,

$$\sum_{\substack{\ell \neq i \\ k \neq j}} \alpha_{\ell k} \cdot x_{\ell k} \leq \sum_{\substack{\ell \neq i \\ k \neq j}} \alpha_{\ell k} \cdot x'_{\ell k}, \tag{2}$$

from optimality of x'.

By (1) and (2) we get that

$$v(P, Q) = \alpha_{ij} + v(P - \{i\}, Q - \{j\}), \quad \text{if } x_{ij} = 1. \tag{*}$$

Note that Lemma 8.15 and (*) together imply that if buyer i gets object j in the auction,

$$\bar{u}_i = \alpha_{ij} - [v(P - \{i\}, Q) - v(P - \{i\}, Q - \{j\})]. \tag{**}$$

That is, buyer i buys object j at the price

$$p_j = [v(P - \{i\}, Q) - v(P - \{i\}, Q - \{j\})].$$

The critical observation for the proof of the next theorem is that this price does not depend on any valuations α_{ik} of buyer i. So as in the Vickrey second-price auction for a single object, the price a buyer pays is not determined by the reserve prices he or she states. This permits us to prove the following.

Theorem 8.16 (Demange; Leonard).   *In the multiobject auction mechanism, truth telling is a dominant strategy for each buyer.*

Proof: If buyer i tells the truth and gets object j at the end of the auction, his or her profit will be $\bar{u}_i = \alpha_{ij} - [v(P - \{i\}, Q) - v(P - \{i\}, Q - \{j\})]$, by (**). Suppose the buyer misrepresents his or her valuations. If he or she is assigned the same object j at the end of the auction under the new valuations, the buyer's true payoff will be the same, since he or she will pay the same price p_j [given by (**)] for object j. If assigned to some other object k, the buyer will pay $[v(P - \{i\}, Q) - v(P - \{i\}, Q - \{k\})]$ and his or her true profit will be $\bar{u}'_i = \alpha_{ik} - [v(P - \{i\}, Q) - v(P - \{i\}, Q - \{k\})]$. But,

$$\alpha_{ik} + v(P - \{i\}, Q - \{k\}) = \alpha_{ik} + \max_{x'} \sum_{\substack{\ell \neq i \\ \ell \neq k}} \alpha_{\ell\ell} \cdot x'_{\ell\ell} \leq v(P, Q)$$

$$= \alpha_{ij} + v(P - \{i\}, Q - \{j\}),$$

by (*). So $\bar{u}_i \geq \bar{u}'_i$, and buyer i has not profited from misstating his or her valuations. If buyer i is unmatched, then he or she also does not profit, since $\bar{u}'_i = 0 \leq \bar{u}_i$.

If buyer i is unmatched under the true valuations, then $v(P-\{i\}, Q) = v(P, Q)$, so if he or she is matched to k under the misstated valuations, $\bar{u}_i' = [\alpha_{ik} + v(P - \{i\}, Q - \{k\})] - v(P, Q) \le v(P, Q) - v(P, Q) = 0 = \bar{u}_i$. Thus in every case, $\bar{u}_i \ge \bar{u}_i'$ and truth telling is a dominant strategy for i.

8.5 The effect of new entrants

In this section we return to another question we have previously considered for the marriage market, namely, What is the effect on the set of stable outcomes of changing the market by introducing a new agent? Aside from being able to prove results parallel to those we have seen for the marriage model, we will see that the special assumptions of the assignment model allow us to draw some even stronger conclusions.

Suppose some P-agent i^* enters the market $M = (P, Q, \alpha)$. The new market is then $M^{i^*} = (P \cup \{i^*\}, Q, \alpha')$, where $\alpha'_{ij} = \alpha_{ij}$ for all i in P and j in Q. The first result, whose proof we will defer until the more general model of the next chapter (Theorem 9.12), is parallel to Theorem 2.25 for the marriage market. It compares the optimal stable outcomes of the two markets.

Proposition 8.17. (a) *Let* $(\bar{u}, \underline{v})$ *and* $(\bar{u}', \underline{v}')$ *be the* P-*optimal stable payoffs for* M *and* M^{i^*}, *respectively. Then* $\bar{u}_i' \le \bar{u}_i$ *for all* i *in* P *and* $\underline{v}_j' \ge \underline{v}_j$ *for all* j *in* Q.

(b) *Let* $(\underline{u}, \bar{v})$ *and* $(\underline{u}', \bar{v}')$ *be the* Q-*optimal stable payoffs for* M *and* M^{i^*}, *respectively. Then* $\underline{u}_i' \le \underline{u}_i$ *for all* i *in* P *and* $\bar{v}_j' \ge \bar{v}_j$ *for all* j *in* Q.

The next result (analogous to Theorem 2.26 for the marriage market) shows that there will be some P- and Q-agents for whom we can unambiguously compare *all* stable outcomes of the two markets.

Theorem 8.18: Strong dominance (Mo). *If* i^* *is matched under some optimal assignment for* M^{i^*}, *then there is a nonempty set* A *of agents in* $P \cup Q$ *such that every* Q-*agent in* A *is better off and every* P-*agent in* A *is worse off at any stable outcome of the new market than at any stable outcome for the old market. That is, for all* (u', v') *and* (u, v) *stable for* M^{i^*} *and* M, *respectively, we have*

(a) *if a* P-*agent* i *is in* A, *then* $u_i \ge u_i'$
(b) *if a* Q-*agent* j *is in* A, *then* $v_j \le v_j'$.

Before proving this theorem we need to recall Lemma 8.15, which implies that if $(\bar{u}', \underline{v}', x')$ is the P-optimal stable outcome for M^{i^*}, then $\bar{u}_{i^*} = v(P \cup \{i^*\}, Q) - v(P, Q)$.

Recall that the central idea of the proof of Lemma 8.15 involved showing that if i^* is assigned by x' to some agent j_1, then there is an oriented path $c = (j_1, i_1, j_2, i_2, \ldots, j_s, i_s, (j_{s+1}))$, starting from j_1, with the following properties.

P1: c ends at i_s if i_s is unassigned by x' or c ends at j_{s+1} if i_s is assigned to j_{s+1} by x' and $\underline{v}'_{s+1} = 0$.

P2: i_m is assigned by x' to j_{m+1} for all $m = 1, \ldots, s-1$.

P3: $\bar{u}'_m + \underline{v}'_m = \alpha_{mm}$ for all $m = 1, \ldots, s$ (since j_m is in the demand set of i_m at prices $\underline{v}'$).

Furthermore if x is the assignment (in M) defined by

(i) $x_{mm} = 1$ for all $m = 1, \ldots, s$,

(ii) $x_{ij} = 1$ if i and j are not in the path and $x'_{ij} = 1$,

(iii)   if j_{s+1} is in the path he or she is unassigned by x,

then x is an optimal assignment for M and the outcome $(\bar{u}, \underline{v}', x)$ is stable for M, where $\bar{u}_i = \bar{u}'_i$ for all i in P. Mo calls the path c a "turnover chain," with the last element being the "crowd-out" i_s or the "draw-in" j_{s+1}. The idea is that if x and x' are the assignments before and after i^* enters the market, then the agents in the chain c are those whose assignments change. If, for example, the P-agent i_s is unassigned by x', he or she has been crowded out of the market by the entry of the new P-agent i^*.

The existence of the path c will be needed to prove Theorem 8.18. The following lemma takes advantage of the special assumptions of the assignment game, namely, that all payoffs are essentially monetary in nature, to compare the benefits and losses that agents in a turnover chain experience when a new player enters the game.

For each i in P and j in Q, define the "benefit functions" B_i and B_j as follows. For all pairs of payoff vectors (u, v) and (u', v'), with (u, v) stable for M and (u', v') stable for M^{i^*},

$$B_i((u, v), (u', v')) = u'_i - u_i, \quad \text{and}$$
$$B_j((u, v), (u', v')) = v'_j - v_j.$$

Lemma 8.19: Benefit lemma (Mo).   *Let x' be an optimal assignment for M^{i^*}. If i^* is matched to some j_1 under x' and $(j_1, i_1, j_2, i_2, \ldots, j_s, i_s, (j_{s+1}))$ is some oriented path satisfying properties* P1, P2, *and* P3, *then*

$$B_{j_1} \geq B_{j_2} \geq \cdots \geq B_{j_s} \geq B_{j_{s+1}}; \quad \text{and}$$
$$B_{i_s} \geq B_{i_{s-1}} \geq \cdots \geq B_{i_1}.$$

The lemma compares the "benefits" that accrue to agents in a turnover chain resulting from the entry of the P-agent i^*. Looking ahead for a moment to when we have completed the proof of Theorem 8.18, we know that these benefits will be nonnegative for all the Q-agents and nonpositive for all the P-agents in the chain. So the lemma says that the greatest benefit will come to agent j_1, who will be matched to i^*, with decreasing benefits to j_2 and so on for Q-agents more distant in the chain from i^*. And the greatest harm (i.e., the most negative benefit) will come to agent i_1, who was matched to j_1 before i^* entered the market, with less harm done to P-agents further down the chain from i^*. Note that these comparisons are meaningful here because we are speaking of monetary gains and losses. (In the marriage model, no similar comparison is possible, since it would involve comparisons of, e.g., how a change from my second to my third choice mate compares with your change from your seventh to your ninth choice.)

Proof of Lemma 8.19:   Let (u', v', x') be stable for M^{i^*} and let (u, v, x) be stable for M, where x is defined from x' by rules (i)–(iii).

Since $x_{11} = 1$, it follows from the stability of (u, v, x) that

$$\alpha_{11} - v_1 \geq \alpha_{12} - v_2. \tag{1}$$

Since $x'_{12} = 1$, the stability of (u', v', x') implies

$$\alpha_{12} - v'_2 \geq \alpha_{11} - v'_1. \tag{2}$$

Adding (1) and (2) gives us that $v'_1 - v_1 \geq v'_2 - v_2$.

In the same manner, from the fact that $x_{22} = 1$ and $x'_{23} = 1$, we obtain

$$v'_2 - v_2 \geq v'_3 - v_3.$$

Repeating this procedure we get that

$$v'_1 - v_1 \geq v'_2 - v_2 \geq \cdots \geq v'_s - v_s \geq v'_{s+1} - v_{s+1}.$$

In an analogous way we obtain

$$u'_1 - u_1 \leq u'_2 - u_2 \leq \cdots \leq u'_s - u_s.$$

Since (u, v) and (u', v') are arbitrary, we have concluded the proof.

Proof of Theorem 8.18:   Consider any path starting from some partner of i^* under some optimal assignment for M^{i^*} and satisfying properties P1, P2, and P3. Let A be the union of all agents belonging to all these such paths. Since i^* is matched under some optimal assignment for M^{i^*}, $A \neq \emptyset$. It is enough to prove the theorem for any such path in A.

Suppose x' is an optimal assignment for M^{i^*} under which i^* is matched to j_1. Let $c = (j_1, i_1, \ldots, j_s, i_s, (j_{s+1}))$ be some oriented path starting from j_1 satisfying properties P1, P2, and P3. Let x be the optimal assignment for M derived from x' by rules (i)–(iii). Let (u', v', x') and (u, v, x) be stable outcomes for M^{i^*} and M, respectively.

Case 1:   The path c ends at i_s. So i_s is unmatched under x'. Then $u'_s = 0$ and so $u_s \ge u'_s$. Since (u, v) and (u', v') are arbitrary, $B_{i_s} \le 0$. From Lemma 8.19 it follows that $B_{i_m} \le 0$ for all $m = 1, 2, \ldots, s-1$. In particular, $u'_m \le u_m$ for all $m = 1, 2, \ldots, s-1$. Now, since $x_{mm} = 1$ for all $m = 1, \ldots, s$, we have $u_m + v_m = \alpha_{mm}$ and $u'_m + v'_m \ge \alpha_{mm}$ by stability, from which it follows that $(v'_m - v_m) + (u'_m - u_m) \ge 0$. We already know that $u'_m - u_m \le 0$. Therefore $v'_m - v_m \ge 0$ for all $m = 1, \ldots, s$, which concludes the proof for this case.

Case 2:   The path c ends at j_{s+1}. So $v'_{s+1} = 0$. Then j_{s+1} is not assigned under x. Hence $v_{s+1} = 0$ and $v'_{s+1} = v_{s+1}$, which implies $B_{j_{s+1}} = 0$. Hence from Lemma 8.19, $B_{j_m} \ge 0$ and in particular, $v'_m - v_m \ge 0$ for all $m = 1, \ldots, s$. As before, since $x'_{s,s+1} = 1$ we obtain that $(v_{s+1} - v'_{s+1}) + (u_s - u'_s) \ge 0$.

Hence we have that $u_s - u'_s \ge 0$. This implies that $B_{i_s} \le 0$, which in turn implies that $B_{i_m} \le 0$ for all $m = 1, \ldots, s$. Then $u'_m - u_m \le 0$ for all $m = 1, \ldots, s$ and the proof is complete.

The final result of this section can be thought of as describing how much the entry of an agent i^* can move the core of the game. There will be some agents whose worst core payoff in one of the two games (with and without i^*) is exactly equal to their best core payoff in the other.

Corollary 8.20 (Mo).   *Let $(\bar{u}', \underline{v}')$ be the P-optimal stable payoff for M^{i^*}. Let $(\underline{u}, \bar{v})$ be the Q-optimal stable payoff for M. If i^* is matched under some optimal assignment for M^{i^*}, there exists a nonempty set A of agents in $P \cup Q$ such that*

(a)   *if a P-agent i is in A, then $\bar{u}'_i = \underline{u}_i$;*
(b)   *if a Q-agent j is in A, then $\underline{v}'_j = \bar{v}_j$.*

Proof: Construct A in the same way as in Theorem 8.18. We know that $(u^*, \underline{v}')$ is a stable payoff for M, where $u_i^* = \bar{u}'_i$, for all i in P. Then, from the Q-optimality of $(\underline{u}, \bar{v})$ it follows that $\underline{v}'_j \le \bar{v}_j$ for all $j \in Q$ and $\bar{u}'_i \ge \underline{u}_i$ for all $i \in P$. Now use Theorem 8.18 (strong dominance) to get

$$\underline{u}_i \ge \bar{u}'_i \ge \underline{u}_i \quad \text{for all } i \text{ in } A,$$

$$\bar{v}_j \le \underline{v}'_j \le \bar{v}_j \quad \text{for all } j \text{ in } A,$$

from which it follows that $\underline{u}_i = \bar{u}'_i$ and $\bar{v}_j = \underline{v}'_j$ for all i and j in A.

8.6   Guide to the literature

The assignment game is a model formulated and studied by Shapley and Shubik (1972). All the initial results presented here are from that paper, although the proofs are not the same.

Section 8.3 follows the paper of Demange, Gale, and Sotomayor (1986). The auction mechanism is a version of the Hungarian algorithm for the assignment problem (see, e.g., Dantzig 1963). Hall's theorem is due to P. Hall (1935). Two simple proofs of Hall's theorem are given by Gale (1960) in a book that deals with some other linear assignment models. Demange, Gale, and Sotomayor (1986) also consider another auction mechanism that is a version of the deferred acceptance algorithm proposed by Crawford and Knoer (1981), which in turn is a special case of the algorithm of Kelso and Crawford that we considered in Section 6.2. They make precise the observation of Crawford and Knoer that the outcome of this algorithm for the discrete case can be made to approximate arbitrarily closely the buyer-optimal core outcome for the continuous assignment problem. They showed that the final price obtained in this algorithm has upper and lower bounds that can be made arbitrarily close to the minimum equilibrium price. Mo (1988b) considers a generalization of the Hungarian algorithm in this context by defining an overdemanded set that contains all minimally overdemanded sets, which he calls the largest pure overdemanded set. Mo, Tsai, and Lin (1988) observe that Demange, Gale, and Sotomayor (1986) incorrectly assert that an algorithm in Gale (1960) computes a minimal overdemanded set, but they show that a variant of this algorithm computes the largest pure overdemanded set.

Section 8.4 follows the independent work of Leonard (1983) and Demange (1982). The proof of Theorem 8.15 presented here follows that of Demange, whereas our proof of Theorem 8.16 follows Leonard's paper.

Section 8.5 on new entrants follows the work of Mo (1988a), although the proofs are somewhat different. Proposition 8.17 will be proved for a generalization of the assignment model in the next chapter. A particular case was also proved (for a different generalization of the assignment model) by Kelso and Crawford (1982). As mentioned in connection with Theorem 2.25, earlier related results in the context of linear programming are found in Shapley (1962). Although most of the results in the literature concern the effect of new entrants on the core of the game, Mo and Gong (1989) show that the same qualitative effects (i.e., agents on the same side of the market are substitutes and agents on opposite sides are complements) are found using the Shapley value. (The Shapley value selects a unique imputation for each game with side payments: See Shapley 1953b, and the collection of papers on the subject in Roth 1988b.)

Becker (1981), who uses the assignment model to study marriage and household economics, makes use of the fact that stable outcomes all correspond to optimal assignments (and that the optimal assignment is typically unique) to study which men are matched to which women, for different assumptions about how the assignment matrix is derived.

Rochford (1984) characterized certain points in the interior of the core of an assignment game as fixed points of a "rebargaining" process, in which matched pairs are thought of as bargaining over their transfer payments. In Roth and Sotomayor (1988) it was observed that Tarski's celebrated fixed point theorem (for order-preserving functions from a complete lattice to itself) implies that these interior fixed points in the core share the lattice property of the core, and have P- and Q-optimal elements. A similar rebargaining process was explored for a generalization of the assignment game by Moldovanu (1988). A different formulation of the bargaining process led Crawford and Rochford (1986) to consider outcomes outside of the core. A similarly motivated reformulation by Bennett (1988), however, again led to points in the core. A strategic model of bargaining and matching that yields core points as equilibria is studied in Kamecke (1989).

Geometric properties of the core of assignment games have also received attention. Some measures of the degree to which the core is "elongated," reflecting the polarization of interests between the two sides of the market, have been considered by Quint (1987a). Balinski and Gale (1987) showed that the number of vertices of the core polytope of the assignment game is at most $\binom{2m}{m}$ where $m = \min\{|P|, |Q|\}$. They gave a characterization of the games that realize these numbers when $|P| = |Q|$ and also studied games where $|P| \neq |Q|$.

A presentation of the assignment game as part of a general introduction to game theory is given by Shubik (1984). A number of generalizations and related models have been explored, for example, by Curiel (1988), Curiel and Tijs (1985), Kaneko (1976, 1982), Kaneko and Wooders (1982), Kaneko and Yamamoto (1986), Kamecke (1987), Quint (1987b, 1988a), and Thompson (1980) (who uses a nonstandard definition of the core, however). Sotomayor (1986b) uses a standard definition of the core for Thompson's model, which allows multiple partners, and observes that this model differs in many respects from the assignment game.

Sotomayor (1988) considers two generalizations of the assignment game that allow many-to-many matching. In a model that keeps track of the individual transactions between each firm and its workers, the results parallel those of Chapter 5. That is, the relationship between the results for her model and the results presented in this chapter and the next for the one-to-one case are very similar to the relationship we have observed between the college admissions model and the marriage model. However when only the aggregate payoffs to each agent are modeled, the core no longer corresponds to the set of pairwise stable outcomes.

Quint (1988b) considers some conditions under which games with more than two "sides" may have nonempty cores.

Samet and Zemel (1984) consider the relationship between linear programs and their duals in connection with the core of side payment games whose coalitional function value for each coalition is given by a linear program. See also Owen (1975) who studies games of this kind.

CHAPTER 9

A generalization of the assignment model

This chapter presents one of the generalizations of the assignment game, in which agents' preferences may be represented by nonlinear utility functions. So in this model agents are allowed to make somewhat more complex tradeoffs than in the assignment model between whom they are matched with and how much money they receive. Nevertheless, each of the principal results we proved for the marriage market has a close parallel in the present model.

This model is a variant of a model introduced by Demange and Gale (1985). The only difference between the model presented here and their model is in the definition of feasible outcomes, which here allow monetary transfers to be made not only among matched pairs of agents, but also among arbitrary coalitions of agents, as in the assignment model and the one-seller model explored in the previous two chapters. Aside from making this model a generalization of these other models, this change allows us not to rule out a priori the kinds of strategic opportunities available to coalitions of bidders, for example, that we discussed in Sections 1.2 and 7.2.1. However most of the results from Demange and Gale's model carry over unchanged to the case when monetary transfers are allowed between unmatched agents. The reason is that as in the assignment game, no such transfers are made at *stable* outcomes. That is, we will see that the only monetary transfers that occur at stable outcomes are between agents who are matched to each other.

Since the results presented for this model have close parallels with those obtained for the marriage model, we will forego detailed description of their intuitive content and instead give a brief indication of which are the most closely related results.

9.1 The model

There are two finite disjoint sets of agents, P and Q, with m and n members, respectively. Members of P are called P-agents and members of Q

are called Q-agents. An outcome of the market matches P-agents with Q-agents. The letters i and j will be reserved to denote P-agents and Q-agents, respectively.

Definition 9.1. *A matching μ is a bijection (a one-to-one correspondence) of $P \cup Q$ onto itself of order two (that is, $\mu^2(x) = x$) and such that if $\mu(i) \neq i$ then $\mu(i)$ is in Q and if $\mu(j) \neq j$ then $\mu(j)$ is in P.*

Any agent who is matched to himself ($\mu(x) = x$) will also be called *unmatched* and an agent will be called *matched* only if he is not matched to himself.

The preferences of the agents are given by utility functions: $u_{ij}(x)$ denotes the utility to i of being matched with j and receiving a monetary payment of x, and $v_{ij}(x)$ denotes the utility to j of being matched to i and receiving a payment of x. We will assume that the functions u_{ij} and v_{ij} are continuous and increasing from R onto R. We will also suppose that for each i and j the utility of being unmatched is given by some numbers $u_{ii}(0) = r_i$ and $v_{jj}(0) = s_j$. (The assignment game is the special case in which all utility functions are linear; for example, in which all r_i and s_j equal 0, and u_{ij} and v_{ij} are of the form $u_{ij}(-x) = \alpha_{ij} - x$ and $v_{ij}(x) = x$.)

Note that we need to make the following strong assumption about the utility functions u_{ij} and v_{ij} for each i in P and j in Q: The range of u_{ij} and v_{ij} is all of R. That is, we assume that each such function is unbounded above and below as a function of the payment x. This implies that (as in the assignment game) a sufficient monetary payment can make any match more desirable than any other match at a given payment. (To see why this is a strong assumption, ask yourself, for example, if there is any amount of money that would make you prefer being single, say, to being married to your current spouse and having a billion dollars.) Since the utility functions are continuous, we are also assuming that no agent has strict preferences.

Let f_{ij} and g_{ij} denote the inverses of u_{ij} and v_{ij}, respectively. We will call them *compensation functions*. So $f_{ij}(u)$ is the amount of money i must receive in order to get the utility level u if i and j are matched, and $f_{ii}(r_i) = g_{jj}(s_j) = 0$. The functions f_{ij} and g_{ij} are also continuous and increasing functions from R onto R. (The fact that the domain of each function is all of R is a consequence of the assumption that the utility functions have all of R as their range.)

For our purposes, it will be more convenient to work with the compensation functions. The sets of f_{ij}, g_{ij}, r_i, and s_j will be denoted by f, g, r, and s, respectively. Then the market is denoted by $M = (P, Q; f, g; r, s)$.

Definition 9.2. *A **feasible payoff** (u, v) of M consists of a vector u in R^m, indexed by the elements of P, and a vector v in R^n, indexed by the elements of Q, such that there exists a matching μ with the property*

$$\sum_{i \in P} f_{i\mu(i)}(u_i) + \sum_{j \in Q} g_{\mu(j)j}(v_j) \le 0.$$

That is, a feasible payoff corresponds to a matching of agents, and possibly to a transfer of monetary payments from some agents to others. But no outside money enters the system: The transfers cannot sum to a positive amount.

The pair (u, v) gives the utility levels for the P- and Q-agents. We say that the matching μ is compatible with (u, v) or that (u, v) has a matching μ. An *outcome* $((u, v); \mu)$ is given by a feasible payoff (u, v) and a compatible matching μ.

Definition 9.3. *A payoff (u, v) is called **pairwise feasible** if there is a matching μ such that, for all P-agents i and Q-agents j, if $\mu(i) = j$ then*

$$f_{ij}(u_i) + g_{ij}(v_j) \le 0 \quad \text{(pairwise affordability);} \quad \text{and}$$
$$u_i = r_i \text{ if } \mu(i) = i, \quad \text{and} \quad v_j = s_j \text{ if } \mu(j) = j.$$

In a pairwise feasible outcome $((u, v); \mu)$, every matched pair and unmatched agent can finance its own utility payoffs without any monetary transfers from other agents. It is clear that every pairwise feasible payoff is also feasible, although the converse is not true.

Definition 9.4. *The feasible payoff (u, v) is **stable** if*

$$u_i \ge r_i, \quad v_j \ge s_j \quad \text{(individual rationality);} \quad \text{and}$$
$$f_{ij}(u_i) + g_{ij}(v_j) \ge 0 \quad \text{for all } (i, j) \text{ in } P \times Q \quad \text{(no blocking pairs).}$$

If the latter inequality did not hold, (u, v) would be blocked by some pair (i, j) in $P \times Q$ because, since f_{ij} and g_{ij} are continuous and increasing, we could get a feasible payoff (u', v') such that $u'_i > u_i$, $v'_j > v_j$ and $f_{ij}(u'_i) + g_{ij}(v'_j) \le 0$. (The latter inequality says that i and j could finance (u'_i, v'_j) without requiring any money from any other players.)

If a matching is compatible with some stable payoff, we will call it a *stable matching*. An outcome for M will be called feasible or stable if the corresponding payoff is feasible or stable.

The following lemma is an immediate consequence of the definitions of feasibility and stability.

Lemma 9.5. *Let $((u, v), \mu)$ be a stable outcome. If $\mu(i) = j$, then*

$$f_{ij}(u_i) + g_{ij}(v_j) = 0.$$

If i is unmatched, then $u_i = r_i$, and $v_j = s_j$ if j is unmatched.

So, as discussed in the introduction to this chapter, stable outcomes are pairwise feasible: At a stable outcome, any monetary transfers occur only between matched pairs of agents.

Define $U_M \equiv \{u \in R^m; (u, v)$ is stable for M for some $v \in R^n\}$ and $V_M \equiv \{v \in R^n; (u, v)$ is stable for M for some $u \in R^m\}$.

It is not difficult to confirm that the set of stable payoffs for M equals the core of M, $C(M)$, when the rules of the market are that any P-agent and Q-agent who both agree may conclude a match, and each agent has the right to remain unmatched.

We will not prove here that stable outcomes exist for this model. Crawford and Knoer (1981) observed that their proof of the existence of stable outcomes in the assignment model did not make use of the linearity of the model, and so could apply to this kind of model also. An explicit proof of the existence of stable outcomes in this model is given by Alkan and Gale (1988), and related existence results will be discussed in Section 9.4.

9.2   The core

In this section we will study the "structural" properties of the set of stable outcomes, that is, of the core of this generalized assignment game. Perhaps the easiest way to summarize the results presented here is as follows. The topological properties of the core in this model are quite different than in the assignment game: Here the core need not be a convex set, and the set of stable outcomes compatible with a given matching need not even be connected (although the whole set of stable outcomes is connected). Nevertheless, the important relationships among stable outcomes (such as the existence of optimal stable outcomes for each side of the market) that we found in the marriage market carry over to the present case, as well as to the assignment game. And in the next section we will see that this is also true of the results concerning the strategic opportunities facing the players.

We begin with an example that illustrates some of the topological properties of the core that we lose when we go from the assignment game to the present model.

Example 9.6: A game with nonconvex core (Roth and Sotomayor)
Let $P = \{p_1, p_2\}$ and $Q = \{q_1, q_2\}$, with reservation prices $r_1 = r_2 = 0$ and $s_1 = s_2 = -4\pi$. The compensation functions are $f_{11}(u) = u - \sin u$, $g_{22}(v) = v + \sin v$, $f_{ij}(u) = u$ if $(i, j) \neq (1, 1)$, $g_{ij}(v) = v$ if $(i, j) \neq (2, 2)$. U_M is not convex. The shape of U_M is shown in Figure 9.1.

Another important property of the core of the assignment game is that every stable matching is compatible with every stable payoff vector.

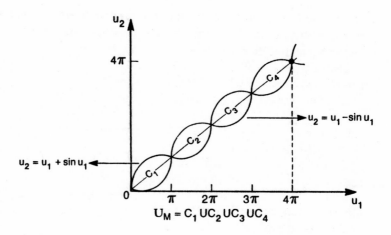

Figure 9.1.

This is no longer true for this example. The payoffs $(u, v) = (\pi/2, \pi/2 + 1;$ $1 - \pi/2, -\pi/2)$ and $(u', v') = (3\pi/2, 1 + 3\pi/2; -1 - 3\pi/2, -3\pi/2)$ are stable with matchings μ and μ', respectively, where $\mu(p_1) = q_1$, $\mu(p_2) = q_2$ and $\mu'(p_1) = q_2$, $\mu'(p_2) = q_1$. However μ is not compatible with (u', v') and μ' is not compatible with (u, v). In Figure 9.1, $C_1 \cup C_3$ corresponds to the stable payoffs for the P-agents that are compatible with μ, and $C_2 \cup C_4$ is the set of payoffs compatible with μ'. As we can see, these are not connected sets, although U_M is connected. This contrasts with the assignment game, in which the set of stable payoffs for the P-agents that are compatible with some stable matching is connected, since, as already noted, it is the whole of U_M.

The following lemma is an analogue of the decomposition lemma (particularly Corollary 2.21) for the marriage problem.

Lemma 9.7: Decomposition (Demange and Gale). *Let $((u, v); \mu)$ and $((u', v'); \mu')$ be stable outcomes for M. Let $P^1 = \{i \in P; u'_i > u_i\}$, $P^2 = \{i \in P; u_i > u'_i\}$ and $P^0 = \{i \in P; u_i = u'_i\}$. Define Q^1, Q^2, and Q^0 analogously. Then $\mu'(P^1) = \mu(P^1) = Q^2$ and $\mu'(P^2) = \mu(P^2) = Q^1$. Furthermore, every agent in P^0 who is matched (under either μ or μ') is matched with an agent in Q^0, and vice versa.*

Lemma 9.7 states that everyone who prefers one of two stable payoffs is matched under both payoffs and is matched with someone who prefers the other payoff.

Proof: All i in P^1 are matched under μ', since $u_i' > u_i \geq r_i$. Analogously, all j in Q^2 are matched by μ, since $v_j > v_j' \geq s_j$. If i is in P^1 then $j = \mu'(i)$ is in Q^2, for if not

$$0 = f_{ij}(u_i') + g_{ij}(v_j') > f_{ij}(u_i) + g_{ij}(v_j),$$

which contradicts the stability of (u, v). On the other hand, if j is in Q^2 then $i = \mu(j)$ is in P^1, for if not

$$0 = f_{ij}(u_i) + g_{ij}(v_j) > f_{ij}(u_i') + g_{ij}(v_j'),$$

which implies instability for (u', v'). Therefore, $\mu'(P^1)$ is contained in Q^2 and Q^2 is contained in $\mu(P^1)$. We have that $|\mu'(P^1)| \leq |Q^2| \leq |\mu(P^1)| = |\mu'(P^1)|$, which implies $\mu(P^1) = \mu'(P^1) = Q^2$. Analogously we prove that $\mu'(P^2) = \mu(P^2) = Q^1$. The last part of the lemma follows from the fact that μ and μ' are one-to-one, and so $\mu(Q^1) = \mu'(Q^1) = P^2$ and $\mu(Q^2) = \mu'(Q^2) = P^1$.

The following theorem is analogous to Theorem 2.22 for the marriage problem. Note two related differences, however. First, Theorem 9.8 does not say that if i is unmatched at some stable outcome then he or she is unmatched at every stable outcome; rather it says that if i is unmatched at some stable outcome then he or she receives the same utility at any stable outcome as if he or she were unmatched, and each stable outcome is compatible with some stable matching at which i is unmatched. Second, whereas the conclusions of Theorem 2.22 depended on the assumption that preferences are strict, in this model the assumption is that everyone is indifferent between different matches at some prices.

Theorem 9.8 (Demange and Gale). *If $((u, v), \mu)$ and $((u', v'), \mu')$ are stable outcomes, then if i (respectively j) is unmatched under μ, $u_i' = r_i$ (respectively $v_j' = s_j$). Furthermore i (respectively j) is unmatched at some stable matching $\bar{\mu}$ that is compatible with (u', v').*

Proof: Suppose i is unmatched under μ. If $u_i' > r_i = u_i$ define P^1 as in Lemma 9.7. Then i is in P^1, so Lemma 9.7 implies i is matched under μ, which is a contradiction. The result for j follows symmetrically. Now, consider (u', v'). We will show that there is a compatible matching $\bar{\mu}$ where i is unmatched. By Lemma 9.7 we may define the matching $\bar{\mu}$ by: $P^1 \cup P^2$ and $Q^1 \cup Q^2$ are matched by μ', P^0 and Q^0 are matched by μ. $\bar{\mu}$ is obviously compatible with (u', v') and matches the same agents as μ. So i is unmatched under $\bar{\mu}$ and the symmetric argument for j completes the proof.

In order to study the core of M as a function of r and s, we will use the notation $M(r, s)$ for the market where r and s may vary but P, Q, f, g are fixed. We have the following parallel to the lattice results (Theorems 2.16, 3.8, and 8.10) for the marriage and assignment models. (Actually Theorem 9.10 is the closest parallel to these previous results, but it will be useful later to have the slightly stronger result stated as follows.)

Lemma 9.9 (Demange and Gale).   *Let $((u, v); \mu)$ be stable for $M(r, s)$ and let $((u', v'); \mu')$ be stable for $M(r', s')$, where $r \leq r' \leq u$ and $s \leq s'$. Then*

(a) $(u \vee u', v \wedge v') \equiv (u^+, v^-)$ *is stable for* $M(r, s)$;
(b) $(u \wedge u', v \vee v') \equiv (u^-, v^+)$ *is stable for* $M(r', s')$.

[*Equivalently, interchanging P and Q, if $r \leq r'$ and $s \leq s' \leq v$, then*

(a) $(u \wedge u', v \vee v')$ *is stable for* $M(r, s)$;
(b) $(u \vee u', v \wedge v')$ *is stable for* $M(r', s')$.]

Proof: Define P^1 and Q^2 as in Lemma 9.7. Since for i in P^1 $u_i' > u_i \geq r_i'$, all of P^1 is matched under μ'. Analogously, since for j in Q^2 $v_j > v_j' \geq s_j' \geq s_j$, all of Q^2 is matched under μ. Using an argument like that in the proof of Lemma 9.7, we obtain that μ and μ' are one-to-one correspondences between P^1 and Q^2. To prove (a) define the matching $\bar{\mu}$ by

P^1 and Q^2 are matched by μ';

$P - P^1$ and $Q - Q^2$ are matched by μ.

We must show that $((u^+, v^-); \bar{\mu})$ is stable for $M(r, s)$. Individual rationality follows since $r \leq r'$ and $s \leq s'$. That there is no blocking pair is also immediate if $u_i^+ = u_i$ and $v_j^- = v_j$ or $u_i^+ = u_i'$ and $v_j^- = v_j'$. If, say, $u_i^+ = u_i$ and $v_j^- = v_j'$, then

$$f_{ij}(u_i^+) + g_{ij}(v_j^-) = f_{ij}(u_i) + g_{ij}(v_j') \geq f_{ij}(u_i') + g_{ij}(v_j') \geq 0$$

by stability of (u', v').

To prove that an unmatched agent gets exactly his individually rational payoff, if i is unmatched by $\bar{\mu}$, then $i \in P - P^1$, so $u_i' \leq u_i = u_i^+$. Furthermore, i is unmatched by μ. Hence $u_i^+ = u_i = r_i$. If j is unmatched by $\bar{\mu}$, then $j \in Q - Q^2$, so $v_j^- = v_j \leq v_j'$ and j is unmatched by μ. Hence $v_j^- = v_j = s_j$. It remains to show that $((u^+, v^-); \bar{\mu})$ is feasible. But this is immediate from the fact that it is pairwise feasible. The pairwise affordability condition follows from the fact that it holds for (u, v) and (u', v').

To prove part (b) of the lemma, define $\underline{\mu}$ by

P^1 and Q^2 are matched by μ;

$P - P^1$ and $Q - Q^2$ are matched by μ'.

The arguments to establish pairwise affordability and the absence of a blocking pair are as before. For individual rationality, $u_i \geq r_i'$ by hypothesis and $u_i' \geq r_i'$ since u' is individually rational, and $v_j^+ \geq v_j' \geq s_j'$. To prove that an unmatched agent gets r_i or s_j, if i is unmatched by μ, then $i \in P - P^1$, so $u_i^- = u_i'$. Furthermore, i is unmatched by μ', so $u_i^- = u_i' = r_i$. If j is unmatched by μ, then $j \in Q - Q^2$, so $v_j' = v_j^+$ and j is unmatched by μ'. Hence $v_j^+ = v_j' = s_j'$.

Theorem 9.10 (Demange and Gale). *The sets U_M and V_M are lattices with smallest and largest elements.*

Proof: The lattice property follows from Lemma 9.9 in the special case where $r' = r$ and $s' = s$. It is easily seen that U_M and V_M are bounded sets:

$$r_i \leq u_i \leq f_{ij}^{-1}(-g_{ij}(s_j))$$
$$s_j \leq v_j \leq g_{ij}^{-1}(-f_{ij}(r_i))$$

if $\mu(i) = j$. Furthermore, it follows from the continuity of f and g that U_M and V_M are closed sets in R^m and R^n, respectively. Hence U_M and V_M are compact lattices and so they have smallest and largest elements.

We will denote by $\bar{u}$ and $\underline{u}$ the largest and smallest elements of U_M, respectively, and by $\bar{v}$ and $\underline{v}$ the largest and smallest elements of V_M, respectively. We will call $(\bar{u}, \underline{v})$ and $(\underline{u}, \bar{v})$ the P-optimal and Q-optimal stable payoffs. The following result is analogous to Theorems 2.24 and 2.25 for the marriage market. (Agents who lower their reservation price are extending their preferences, although if the reservation price is lowered from a very high level the agent can be viewed as entering the market (Theorem 9.12).)

Proposition 9.11 (Demange and Gale). *If $(\bar{u}(r, s), \underline{v}(r, s))$ is the P-optimal stable payoff in $M(r, s)$, then $\bar{u}(r, s)$ is an increasing function of r and a decreasing function of s, and $\underline{v}(r, s)$ is a decreasing function of r and an increasing function of s.*

Proof: Suppose $r' \geq r$ and (u', v') is stable for $M(r', s)$. Then from part (b) of Lemma 9.9 (interchanging P and Q and taking $s' = s$), we have $(u' \vee \bar{u}(r, s), v' \wedge \underline{v}(r, s))$ is stable for $M(r', s)$, so $\bar{u}(r', s) \geq \bar{u}(r, s)$ and $\underline{v}(r', s) \leq \underline{v}(r, s)$. Suppose $s' \geq s$ and (u, v) is stable for $M(r, s)$. Then from part (a) of Lemma 9.9, taking $r = r'$, $(u \vee \bar{u}(r, s'), v \wedge \underline{v}(r, s'))$ is stable for $M(r, s)$, so $\bar{u}(r, s) \geq \bar{u}(r, s')$ and $\underline{v}(r, s) \leq \underline{v}(r, s')$.

The following special case of Proposition 9.11 makes clear the connection with new entrants to the market.

Theorem 9.12 (Demange and Gale).   *If additional Q-agents enter the market, then $\bar{u}$ does not decrease and $\underline{v}$ does not increase. If additional P-agents enter the market, then $\bar{u}$ does not increase and $\underline{v}$ does not decrease.*

Proof: Consider the market $M(P, Q'; f', g'; r, s')$, where Q is contained in Q', $g'_{ij} = g_{ij}$, $s'_j = s_j$ and $f'_{ij} = f_{ij}$ for all i in P and j in Q. We will show that $\bar{u}_i(r, s) \leq \bar{u}_i(r, s')$ for all i in P, and $\underline{v}_j(r, s) \geq \underline{v}_j(r, s')$ for all j in Q. In fact, construct a new market M'' by choosing $s''_j = s'_j$ in Q and for all j in $Q' - Q$, $s''_j > s'_j$ and $g'_{ij}(s''_j) > -f'_{ij}(r_i)$ for all i in P. This means that j will never be matched by any stable matching for M'', for all j in $Q' - Q$. Then $\bar{u}_i(r, s) = \bar{u}_i(r, s'')$ for all i in P and $\underline{v}_j(r, s) = \underline{v}_j(r, s'')$ for all j in Q. From Proposition 9.11 it follows that $\bar{u}_i(r, s) = \bar{u}_i(r, s'') \leq \bar{u}_i(r, s')$ for all i in P and $\underline{v}_j(r, s) = \underline{v}_j(r, s'') \geq \underline{v}_j(r, s')$, for all j in Q. A symmetrical argument applies for the other assertion.

9.2.1   *Some technical results*

Corollary 9.13 through Proposition 9.17 are technical results that will be useful in what follows, but which the reader only interested in the principle results might omit on first reading.

Corollary 9.13.   *If $s \leq s' \leq \bar{v}$ then every stable payoff for $M(r, s')$ is stable for $M(r, s)$.*

Proof: It is clear that $(\underline{u}, \bar{v})$ is stable for $M(r, \bar{v})$. We claim that $(\underline{u}, \bar{v})$ is the Q-optimal payoff for $M(r, \bar{v})$. In fact, let (u, v) be stable for $M(r, \bar{v})$. From Lemma 9.9(a) (interchanging P and Q and taking $r' = r$, $s' = \bar{v}$) we have that $(u \wedge \underline{u}, v \vee \bar{v})$ is stable for $M(r, s)$. Hence $u \geq \underline{u}$ and $v \leq \bar{v}$ and $(\underline{u}, \bar{v})$ is the Q-optimal payoff for $M(r, \bar{v})$. From Proposition 9.11 and the fact that $\bar{v} \geq s'$ it follows that

$$\underline{u}(r, s') \geq \underline{u} \quad \text{and} \quad \bar{v}(r, s') \leq \bar{v}. \tag{1}$$

Now, let (u', v') be a stable payoff for $M(r, s')$. Then,

$$u' \geq \underline{u}(r, s') \quad \text{and} \quad v' \leq \bar{v}(r, s'). \tag{2}$$

By Lemma 9.9(a), (1) and (2) above, and the fact that $s \leq s'$, we have that

$$(u', v') = (u' \vee \underline{u}, v' \wedge \bar{v}) \quad \text{is stable for } M(r, s),$$

and we have completed the proof.

Lemma 9.14 (Demange and Gale). *If $|P| \leq |Q|$ then $\underline{v}_j = s_j$ for some j in Q.*

Proof: The proof is immediate unless $|P| = |Q|$ and all of P is matched under $(\bar{u}, \underline{v})$. Suppose that even in this case $\underline{v}_j > s_j$ for all j in Q. It is clear that $(\bar{u}, \underline{v})$ is stable for $M(\bar{u}, s)$. It follows from Corollary 9.13 (interchanging P and Q) that $(\bar{u}, \underline{v})$ remains a P-optimal payoff for $M(\bar{u}, s)$. Now consider the market $M(r^n, s)$, where $r^n = \bar{u} + 1/n$. Then, by Proposition 9.11 we have that

$$\bar{u}(r^n, s) \geq \bar{u}(r^{n+1}, s) \geq \bar{u} \quad \text{and} \quad \underline{v}(r^n, s) \leq \underline{v}(r^{n+1}, s) \leq \underline{v} \qquad (1)$$

for all n. Hence, the sequence $(\bar{u}(r^n, s), \underline{v}(r^n, s))$ is convergent, and let (u^*, v^*) be its limit when n tends to ∞. Since there is a finite number of matchings, there is some matching μ^*, which is compatible to infinitely many terms of the sequence $(\bar{u}(r^n, s), \underline{v}(r^n, s))$. From continuity of f and g it follows that μ^* is compatible with (u^*, v^*). By (1)

$$u^* \geq \bar{u} \quad \text{and} \quad v^* \leq \underline{v}. \qquad (2)$$

So (u^*, v^*) is stable in $M(\bar{u}, s)$. By the P-optimality of $(\bar{u}, \underline{v})$,

$$u^* \leq \bar{u} \quad \text{and} \quad v^* \geq \underline{v}. \qquad (3)$$

From (2) and (3) it follows that $u^* = \bar{u}$ and $v^* = \underline{v}$. Hence $v_j^* > s_j$ for all j in Q, so for some N sufficiently large $\underline{v}_j(r^N, s) > s_j$ for all j in Q. Hence all of Q must be matched and all of P is matched under $(\bar{u}(r^N, s), \underline{v}(r^N, s))$, so $(\bar{u}(r^N, s), \underline{v}(r^N, s))$ is stable for $M(\bar{u}, s)$. But $\bar{u}_i(r^N, s) \geq r_i^N > \bar{u}_i$, for all i in P, which contradicts P-optimality of $(\bar{u}, \underline{v})$ for $M(\bar{u}, s)$.

Lemma 9.15 (Demange and Gale). *Suppose $\underline{v}_j > s_j$ for all $j \in Q'$ contained in Q and let $P' = \mu(Q')$, where μ is some compatible matching with $(\bar{u}, \underline{v})$. Then there is a pair (i, j) with $f_{ij}(\bar{u}_i) + g_{ij}(\underline{v}_j) = 0$, and $i \in P - P'$ and $j \in Q'$.*

Proof: Note from Lemma 9.14, $P - P' \neq \emptyset$. Arguing by contradiction, suppose $f_{ij}(\bar{u}_i) + g_{ij}(\underline{v}_j) > 0$ for all $i \in P - P'$, $j \in Q'$. Then for some positive λ, $\underline{v}_j - \lambda > s_j$ and

$$f_{ij}(\bar{u}_i) + g_{ij}(\underline{v}_j - \lambda) > 0 \quad \text{for } i \in P - P', j \in Q'. \qquad (1)$$

Let $M' = (P', Q'; f, g; r, s')$ where $s_j' = \underline{v}_j - \lambda$, and let $(\bar{u}', \underline{v}')$ be the P-optimal payoff for M'. By Lemma 9.14, $\underline{v}_k' = \underline{v}_k - \lambda$ for some k in Q'. We claim the payoff (u^*, v^*) is stable for the original market, where

$$u_i^* = \bar{u}_i \text{ for } i \in P - P'$$
$$= \bar{u}_i' \text{ for } i \in P';$$

$$v_j^* = \underline{v}_j' \text{ for } j \in Q'$$
$$= \underline{v}_j \text{ for } j \in Q - Q'.$$

In fact, the only possible unstable pairs (i, j) must have either $i \in P - P'$, $j \in Q'$, or $i \in P'$, $j \in Q - Q'$. In the first case it is not possible because of (1). In the second case note that the restriction of $(\bar{u}, \underline{v})$ to Q' and P' is stable for M', so $\bar{u}_i' \geq \bar{u}_i$ for all $i \in P'$. Then

$$f_{ij}(\bar{u}_i') + g_{ij}(\underline{v}_j) \geq f_{ij}(\bar{u}_i) + g_{ij}(\underline{v}_j) \geq 0,$$

by stability of $(\bar{u}, \underline{v})$. However, $v_k^* < \underline{v}_k$, which contradicts Q-minimality of $(\bar{u}, \underline{v})$.

For the sake of simplicity we will use the notation $M(s)$ for the market where only s may vary.

Lemma 9.16 (Gale).   *Let $v < v'$ in V_M and let $Q' = \{j \in Q; \; v_j' > v_j\}$. Choose $v < s^* < v'$ so that $v_j < s_j^* < v_j'$ for j in Q'. Then there is some (u^*, v^*) stable for $M(s^*)$ such that $v_j^* = s_j^*$ for some j in Q' and $v^* < v'$.*

Proof: Let μ and μ' be compatible matchings for v and v', respectively. By Lemma 9.7 $\mu(Q') = \mu'(Q')$. Let $P' = \mu'(Q')$ and let $(\bar{u}'', \underline{v}'')$ be the P-optimal payoff for $M' = (P', Q', f, g, r, s^*)$. Now define (u^*, v^*) to agree with $(\bar{u}'', \underline{v}'')$ on P' and Q' and with (u, v) on $P - P'$ and $Q - Q'$ and let μ^* be the corresponding matching. By Lemma 9.14 $\underline{v}_j'' = s_j^*$ for some j in Q'. We will show that (u^*, v^*) is stable for $M(s^*)$, which proves the first assertion. It is clear that (u^*, v^*) is feasible. Consider a possible blocking pair (i, k). If $k \in Q'$ and $i \in P - P'$ then $v_k^* \geq s_k^* > v_k$ so if (i, k) blocks (u^*, v^*) we will have

$$0 > f_{ik}(u_i^*) + g_{ik}(v_k^*) > f_{ik}(u_i) + g_{ik}(v_k)$$

which contradicts stability of (u, v). Now observe that (u', v') is stable for M' when restricted to P' and Q'. Then for $i \in P'$ and $k \in Q'$ we have

$$u_i^* = \bar{u}_i'' \geq u_i' \quad \text{and} \quad v_k^* \leq v_k'. \tag{1}$$

Hence if k is in $Q - Q'$, i is in P' and (i, k) blocks (u^*, v^*) we will have

$$0 > f_{ik}(u_i^*) + g_{ik}(v_k^*) = f_{ik}(\bar{u}_i'') + g_{ik}(v_k) \geq f_{ik}(u_i') + g_{ik}(v_k')$$

which contradicts stability of (u', v'). The second assertion, $v^* < v'$, follows from the fact that $v_k^* = \underline{v}_k'' \leq v_k'$, for $k \in Q'$, by (1); $v_k^* = v_k = v_k'$ for $k \in Q - Q'$, by hypothesis and $v_j^* = s_j^* < v_j'$. Thus the proof is complete.

Proposition 9.17 (Gale; Sotomayor).   *The function $\underline{v}(s')$ is continuous for all s' in $[s, \bar{v}] = [s_1, \bar{v}_1] \times \cdots \times [s_n, \bar{v}]$.*

Proof: By Corollary 9.13 $\underline{v}(s') \in V_M$ for all $s' \in [s, \bar{v}]$. If $\underline{v}(s')$ were discontinuous at s', since V_M is compact, there would be a sequence $s^n \to s'$ as $n \to \infty$, s^n in $[s, \bar{v}]$, such that

$$\underline{v}(s^n) \to v' \neq \underline{v}(s') \tag{1}$$

and $v' \in V_M$. Since $\underline{v}(s^n) \geq s^n$, it follows that $v' \geq s'$, so

$$v' = \underline{v}(v') \geq \underline{v}(s') \tag{2}$$

using the monotonicity of the function $\underline{v}$.

From (1) and (2) it follows that $v' > \underline{v}(s')$. Let Q' and s^* be as in Lemma 9.16, that is, $\underline{v}_j(s') < s_j^* < v_j'$ for all j in Q' and $\underline{v}_j(s') = s_j^* = v_j'$ for all j in $Q - Q'$. Then, for n sufficiently large we must have

(a) $s^n < s^*$, since $s^n \to s'$ as $n \to \infty$ and $s' \leq \underline{v}(s') < s^*$; and

(b) $\underline{v}_j(s^n) > s_j^*$, for j in Q', since $\underline{v}(s^n) \to v'$ as $n \to \infty$, and $v_j' > s_j^*$.

But from Lemma 9.16 there is a stable payoff (u^*, v^*) for $M(s^*)$ with

$$v_j^* = s_j^* \tag{3}$$

for some j in Q'. Every agent in Q' is matched under (u^*, v^*), so if $k \in Q$ is unmatched under μ^*, it follows that k is in $Q - Q'$, so it is unmatched under μ. By Theorem 9.8 this means that $\underline{v}_k = \bar{v}_k = s_k$, which implies that $s_k^n = s_k = s_k^*$, since s^n is in $[s, \bar{v}]$. Then $v_k^* = s_k^n$. Since, from (a), $s^n < s^*$, we can conclude that (u^*, v^*) is stable for $M(s^n)$, so $s_j^* = v_j^* \geq \underline{v}_j(s^n)$, by (3) and the Q-minimality of $(\bar{u}(s^n), \underline{v}(s^n))$. But this contradicts (b) and completes the proof.

9.2.2 The structure of the core

Theorem 9.18 (Gale). *The core is connected.*

Proof: Denote the core of market M by C. Let $S \cup S' = C$ where S and S' are disjoint sets and S' is closed and $(\bar{u}, \underline{v}) \in S$. Let (u', v') be a Q-minimal element of S', that is, there is no (u, v) in S' with $v < v'$. We will show that (u', v') is in the closure of S, so C is connected.

Since $v' \neq \underline{v}$ let $Q' - Q$ be all j such that $v_j' > \underline{v}_j$, and let $P' = \mu(Q')$, where μ is compatible with (u', v'). Now choose any s^* such that $\underline{v}_j < s_j^* < v_j'$ for j in Q' and $s_j^* = v_j' = v_j$ otherwise. By Lemma 9.16 there exists some (u^*, v^*) stable for $M(s^*)$ such that $v^* < v'$ and $v_j^* = s_j^*$ for some j in Q'. Furthermore, from Corollary 9.13 it follows that (u^*, v^*) is stable for M. Then $(u^*, v^*) \in S$ by the Q-minimality property of (u', v'). Thus any neighborhood of (u', v') contains points of S, so (u', v') is in the closure of S.

Theorem 9.18 can also be proved by making use of the continuity of the function $v(s)$.

The following result is the analogue of Theorem 2.27 for the marriage market.

Theorem 9.19: Pareto optimality (Demange and Gale).   *Let $(\bar{u}, \underline{v})$ be the P-optimal payoff and let (u, v) be any Q-individually rational pairwise feasible payoff. Then it is not the case that $u_i > \bar{u}_i$ for all $i \in P$.*

Proof: Let μ be a matching corresponding to (u, v). If $u_i > \bar{u}_i$ for all $i \in P$, then $u_i > r_i$ for all $i \in P$, so all of P is matched by μ. Also $v_j < \underline{v}_j$ for j in $\mu(P)$, for if $v_j \geq \underline{v}_j$ for some j in $\mu(P)$, say, $j = \mu(i)$,

$$0 \geq f_{ij}(u_i) + g_{ij}(v_j) > f_{ij}(\bar{u}_i) + g_{ij}(\underline{v}_j)$$

which contradicts stability of $(\bar{u}, \underline{v})$. Hence $\underline{v}_j > s_j$ for $|P|$ elements of Q, so all of P is matched under $(\bar{u}, \underline{v})$ by some matching $\bar{\mu}$. However, $(\bar{u}, \underline{v})$ is still the P-optimal payoff for $M' = (P, \bar{\mu}(P); f, g; r, s)$. Then by Lemma 9.14, we have $\underline{v}_j = s_j$ for some j in $\bar{\mu}(P)$. If $j \in \mu(P)$ we cannot have $v_j < \underline{v}_j$ by Q-rationality of (u, v), which is a contradiction. Otherwise, j is unmatched under μ, so there is some $k \in \mu(P)$ such that k is unmatched under $\bar{\mu}$. This means that $v_k < \underline{v}_k = s_k$, which contradicts individual rationality of (u, v).

The analogue of Lemma 3.5 for the marriage market is the following.

Lemma 9.20: Blocking lemma (Demange and Gale).   *Let $((u, v); \mu)$ be a feasible pairwise outcome for M and let $P^+ = \{i \in P; u_i > \bar{u}_i\}$. Then there are an i in $P - P^+$ and a j in $\mu(P^+)$ such that $f_{ij}(u_i) + g_{ij}(v_j) < 0$.*

Proof: Let $\bar{\mu}$ be a matching compatible with $(\bar{u}, \underline{v})$. There are two cases.

Case 1: $\mu(P^+) \neq \bar{\mu}(P^+)$. Since $u_i > \bar{u}_i \geq r_i$ for i in P^+, it follows that P^+ is matched by μ. Choose j in $\mu(P^+)$, $j \notin \bar{\mu}(P^+)$, say, $j = \mu(p)$. Since $u_p > \bar{u}_p$, it follows that $v_j < \underline{v}_j$, for if not,

$$0 \geq f_{pj}(u_p) + g_{pj}(v_j) > f_{pj}(\bar{u}_p) + g_{pj}(\underline{v}_j)$$

which contradicts stability of $(\bar{u}, \underline{v})$. Hence j is matched under $\bar{\mu}$, say, $j = \bar{\mu}(i)$, where $i \in P - P^+$. By feasibility, $f_{ij}(\bar{u}_i) + g_{ij}(\underline{v}_j) \leq 0$ but $u_i \leq \bar{u}_i$ since $i \in P - P^+$, and $v_j < \underline{v}_j$, so the assertion is proved since f and g are increasing.

Case 2: $\mu(P^+) = \bar{\mu}(P^+)$. Following the proof of Case 1, the fact that $u_i > \bar{u}_i$ for all i in P^+ implies that $s_j \leq v_j < \underline{v}_j$ for all j in $\mu(P^+)$. By Lemma 9.15, there is i in $P - P^+$ and j in $\mu(P^+)$ such that $f_{ij}(\bar{u}_i) + g_{ij}(\underline{v}_j) = 0$ and the result follows since $v_j < \underline{v}_j$ and $u_i \leq \bar{u}_i$.

Remark 1:   Since $i \in P - P^+$, $u_i \leq \bar{u}_i$. Since $j \in \mu(P^+)$, there exists some $p \in P^+$ such that $\mu(p) = j$. Then $u_p > \bar{u}_p$, so $v_j < \underline{v}_j \leq \bar{v}_j$ by stability of $(\bar{u}, \underline{v})$.

The following result is precisely analogous to Theorem 3.4 for the marriage model.

Theorem 9.21: Strong stability (Sotomayor). *Let $((u,v);\mu)$ be a feasible pairwise outcome not stable for $M = (P, Q, f, g, r, s)$. Then either there exists a pair (i, j) such that*

$$f_{ij}(u_i) + g_{ij}(v_j) < 0 \tag{1}$$

and a stable payoff (u', v') such that

$$u'_i \geq u_i \quad and \quad v'_j \geq v_j \tag{2}$$

or $((u, v); \mu)$ is not individually rational.

Note that if (u, v) is not individually rational, some agent, say i, has a payoff $u_i < r_i$, so $u'_i > u_i$ is true for any stable (u', v').

Proof of Theorem 9.21: Assume that (u, v) is individually rational. By Lemma 9.20 we need only consider the case where

$$\bar{u} \geq u \quad and \quad \bar{v} \geq v.$$

In fact, if for example, $\bar{u}_i < u_i$ for some $i \in P$, the set P^+ defined in Lemma 9.20 would be nonempty and the pair (i, j) given by the lemma would satisfy the remark following Lemma 9.20, so (1) and (2) would be true with (i, j) and $(u', v') = (\bar{u}, \underline{v})$.

Now consider the market $M^* = (P, Q; f, g; r, v)$. Let $((u^*, v^*); \mu^*)$ be the P-optimal outcome for M^*. From Corollary 9.13, it follows that (u^*, v^*) is stable for M, so $v^* \geq v$.

Now we have three cases.

Case 1: $u_i^* > u_i$ for some i. Then i is matched to some j under μ^*. Since $v_j^* \geq v_j$ we have that (i, j) and (u^*, v^*) satisfy (1) and (2).

Case 2: $u_i^* < u_i$ for some i. Then the blocking lemma applies to M^*, since $P^+ \neq \emptyset$. Then there is a pair (i, j) such that $f_{ij}(u_i) + g_{ij}(v_j) < 0$, $u_i \leq u_i^*$, and $v_j \leq v_j^*$. Hence (i, j) and (u^*, v^*) satisfy (1) and (2).

Case 3: $u^* = u$. Since (u, v) is unstable for M this means that there exists some j such that $v_j^* > v_j$. Then j must be matched to some i under μ^* and clearly (i, j) and (u^*, v^*) satisfy (1) and (2), and the proof is complete.

9.3   Incentives

In this section we consider the incentives facing the agents in the strategic game that arises when a stable matching mechanism is used. By a stable matching mechanism for this model we mean a function that for any P, Q

and any stated preferences (f, g, r, s) produces a stable allocation for the market $M = (P, Q; f, g; r, s)$.

We know from Theorem 7.3 and the remarks following it that no such mechanism exists for which stating the true preferences is a dominant strategy for all agents.

Let $(P, Q; f, g; r, s)$ be some market. If $\{f', r'\}$ and $\{g', s'\}$ are the selected strategies and $((u, v); \mu)$ is the stable outcome for $(P, Q; f', g'; r', s')$ given by the mechanism, then the transfers to P-agents are $f'_{ij}(u_i)$ if $\mu(i) = j$ and 0 if i is unmatched, and the transfers to Q-agents are $g'_{ij}(v_j)$ if $\mu(i) = j$ and 0 if j is unmatched. The utility of each player is of course computed from the matching he or she makes and monetary transfer he or she receives in terms of his or her *true* utility function. Therefore, the *true payoff* under $((u, v); \mu)$ is

$$u_i^* = f_{ij}^{-1}(f'_{ij}(u_i)) \quad \text{if } \mu(i) = j$$
$$= r_i \quad \text{if } i \text{ is unmatched,}$$
$$v_j^* = g_{ij}^{-1}(g'_{ij}(v_j)) \quad \text{if } \mu(j) = i$$
$$= s_j \quad \text{if } j \text{ is unmatched.}$$

Note that we are looking for the moment at the outcome that results directly from the stable matching mechanism. Only pairwise feasible monetary transfers have been considered.

The following theorem is analogous to Theorem 4.10 for the marriage market.

Theorem 9.22 (Demange and Gale).   *Let $((u', v'); \mu)$ be any stable outcome for the market $M' = (P, Q, f', g, r', s)$ and let P' be the set of agents who misrepresented their preferences. Let $(\bar{u}, \underline{v})$ be the P-optimal stable payoff for $M = (P, Q, f, g, r, s)$ and let (u^*, v^*) be the true payoff under $((u', v'); \mu)$. Then $\bar{u}_i \geq u_i^*$ for at least one i in P'.*

Theorem 9.22 is an immediate consequence of the following result, which is the analogue of Theorem 4.11 for the marriage model.

Theorem 9.23 (Sotomayor).   *Let $((u', v'); \mu)$ be any stable outcome for the market $M' = (P, Q, f', g', r', s')$. Let $P' \cup Q'$ be the set of agents who misrepresented their preferences. Let (u^*, v^*) be the true payoff under $((u', v'); \mu)$. Then, there exists a stable payoff (u, v) for the original market such that $u_i \geq u_i^*$ for at least one i in P' or $v_j \geq v_j^*$ for at least one j in Q'.*

Proof: Suppose that all agents in $P' \cup Q'$ are *strictly* better off at some stable outcome $((u', v'); \mu)$ for M' than at *any* stable outcome under the true preferences.

Denote by (u^*, v^*) the true payoff under the outcome $((u', v'); \mu)$. Then

$$u_i^* > \bar{u}_i \quad \text{for all } i \in P', \tag{1}$$
$$v_j^* > \bar{v}_j \quad \text{for all } j \in Q'.$$

The outcome $((u^*, v^*); \mu)$ is clearly pairwise feasible for the original market, since $f_{ij}(u_i^*) + g_{ij}(v_j^*) = f_{ij}'(u_i') + g_{ij}'(v_j') \le 0$, if $\mu(i) = j$, by pairwise feasibility of (u', v') in M'. If $P' \ne \emptyset$, there exists a pair (i, j) in $P \times Q$ such that $f_{ij}(u_i^*) + g_{ij}(v_j^*) < 0$ by the blocking lemma, and such that $\bar{u}_i \ge u_i^*$ and $\bar{v}_j \ge v_j^*$ by the remark following that lemma.

This means that $(i, j) \notin P' \cup Q'$ and hence $u_i^* = u_i'$ and $v_j^* = v_j'$; $f_{ij}' = f_{ij}$ and $g_{ij}' = g_{ij}$. But then, $f_{ij}'(u_i') + g_{ij}'(v_j') < 0$, which contradicts the stability of (u', v') in M'. If $P' = \emptyset$, Q' is nonempty, and the symmetric argument applies.

Observe that Theorem 9.22 does *not* imply that a coalition of P-agents cannot manipulate the P-optimal stable mechanism, since monetary transfers within such a coalition are possible. This is precisely the phenomenon discussed in Section 1.2 and modeled in Section 7.2.1 for a special case of the model we consider in this chapter.

Note however that Theorem 9.22 does imply the following analogue to Theorem 4.7 for the marriage market.

Theorem 9.24. *The optimal stable mechanism for one side of the market makes it a dominant strategy for the agents on that side of the market to state their true preferences.*

So we can summarize as follows:

Proposition 9.25. *When the optimal stable mechanism for one side of the market is employed, in order for coalitions of agents from that side to all profit by misstating their preferences, they must be able to arrange transfers from those who directly profit to those who do not.*

We already know from Theorem 7.3 and its proof, for the special case of the one-seller assignment game, that any stable mechanism will typically give some agent an incentive to misrepresent his or her preferences. That is, in the present model we can prove results analogous to Theorems 4.4 and 4.6 for the marriage market. It is also possible to prove theorems about strategic equilibria that closely parallel Theorems 4.16 and 4.17, for example. One difference between the equilibrium results for this model and for the marriage model is an observation due to Sotomayor (1986a) that although every strategic equilibrium corresponds to a stable outcome, not every stable payoff can be achieved by a strategic equilibrium. However, as in the case of the marriage model, the equilibrium results depend heavily on the assumption that the agents know one another's preferences, and so, as we found in Section 4.5, these results are not robust to relaxations of the assumption of complete information.

9.4   Guide to the literature

The model presented here was introduced by Demange and Gale (1985), except that they defined an outcome to be feasible only if it was what we call pairwise feasible. Thus their model did not permit any side payments between agents other than matched pairs. We have not ruled out side payments, in order that the model presented here should be a generalization of the assignment game, and in order not to inadvertently exclude from consideration in this way the very real strategic possibilities available to coalitions. Nevertheless, as remarked in the introduction, most of the results of their paper require no change because at stable outcomes the only transfers that occur are between matched pairs. The proofs of their results presented here therefore closely follow those in their paper.

In Section 9.2, Example 9.6 is from Roth and Sotomayor (1988). Except for Lemma 9.16, Proposition 9.17, and Theorems 9.18 and 9.19, the other results are from Demange and Gale's paper. Lemma 9.16 is due to Gale. Proposition 9.17, which asserts that the function $v(s)$ is continuous, was proved independently by Gale and by Sotomayor. Gale's proof is simpler and shorter than Sotomayor's proof and it is the proof we presented here. Both results, with Gale's proof for Proposition 9.17, appear in Roth and Sotomayor (1988). The connectivity of the core was also proved independently by Gale and by Sotomayor and appears in Sotomayor (1987). Theorem 9.21 was proved by Sotomayor (1987).

In Section 9.3, Theorem 9.22 is from Demange and Gale (1985). Results analogous to those in Section 4.4.2, concerning the equilibrium strategies when a P-optimal stable mechanism is employed, are found in Demange and Gale (1985) and Sotomayor (1986a).

Roth and Sotomayor (1988) examine a family of interior points in the core of this model, following the work of Rochford (1984) referred to in the previous chapter.

As remarked earlier, Crawford and Knoer (1981) observed that their proof of the existence of stable outcomes in the assignment model did not make use of the linearity of the model, and so could apply to this kind of model also. An explicit proof of existence of stable outcomes was given by Quinzii (1984). A short proof of a related existence theorem was given by Gale (1984), who uses a generalization of the lemma of Knaster,

Kuratowski, and Mazurkiewicz. A proof by Alkan and Gale (1988) that stable outcomes exist depends only on linear programming and duality results for the assignment problem. Alkan (1988a) uses this proof as the basis for an algorithm that finds a point in the core in finitely many steps if utility functions are piecewise linear, and that approximates a core point for general continuous utilities. Alkan (1988b) considers a multiitem auction mechanism for a related model.

PART IV

Epilogue

It seems appropriate to end this volume with a few words about why we have selected and organized the material as we have, and what we think has been learned. To put it another way, now is a good time to explain the title of the book, which includes both "modeling" and "analysis." The fact that the book contains many theorems should suggest what the analysis consists of. And the fact that we began the book with a detailed description of a particular labor market and with a brief description of some auction phenomena, should suggest at least part of what we mean by modeling. But if that were all there was to it, we could have finished our work in many fewer chapters.

Instead, we analyzed a whole family of closely related models, discrete and continuous, with and without complete information, with and without money, with firms employing one worker or many, and with simple or complex preferences. One purpose of these final remarks is to make clear what we think the consideration of all these models together adds to our understanding and interpretation of each of them.

The essence of making and analyzing mathematical models is to distill out the essential features of the situation being modeled, and use them to gain insight into the important properties of the situation. But it is no easy task to discern which features of a situation are essential, and which properties of a model are truly insights into the situation that motivated it.

Of course, one way of testing the quality of an insight is empirically, by considering its ability to predict and organize what we observe. There is certainly no substitute for this. The ultimate test of the practical contribution of this kind of theory will lie in whether it can be used to understand other kinds of markets and two-sided matching situations, and how their performance is related to their organization, as it has so far been used to understand markets like those for new physicians in the United States and the United Kingdom. There is no question that the present status of this body of theory, and the most pressing directions for refining and extending it, derive in large part from this kind of empirical work.

But by considering families of closely related models, as we have done in this volume, we can learn which conclusions are robust to which changes in the models. In this way we can also begin to sort out which properties of the models have a chance of being insights, and which might be consequences of assumptions introduced into (or left out of) the model just for simplicity or tractability.

For example, we saw across a range of models that it may be possible to organize a two-sided matching market in such a way that it is a dominant strategy for all the agents on one side of the market to behave straightforwardly. This means that those agents cannot strategically manipulate the market. But if we had only looked at the marriage model, we might have been led to conclude that even coalitions of these agents could not manipulate the market. However we have seen that this latter conclusion is not robust to any of a number of small changes we might introduce into the model, but is an artifact of the details, and not the essentials, of this initial model. By analyzing a family of models, we were led to reinterpret this result to mean that in order for such coalitions to profitably manipulate the outcome, there must be some way for them to distribute the benefits among themselves. Similarly, if we had considered only models of complete information, we might not have detected how much more robust the dominant strategy results are than the equilibrium results. And if we had assumed that models of one-to-one matching were good proxies for situations involving many-to-one matching, we would not have discovered how even the dominant strategy results are less robust than the existence of stable matchings or the nonexistence of strategy-proof stable mechanisms.

At the risk of belaboring the point (about which, it seems to us, there has often been confusion in the literature), when it comes to interpreting mathematical models, not all theorems are created equal. Although they all may be true consequences of the assumptions of some model, they may not be equally reliable instruments for understanding the world. And the failure to appreciate this can be a two-edged sword, which causes the mathematically inclined to draw erroneous conclusions, and the empirically inclined to reject useful models because of such conclusions.

Examining a range of related models allows us to do a kind of "conceptual sensitivity analysis" on the assumptions of each model. This gives us a chance to assess which features of the models are most important for which kinds of results, and thus to consider the range of situations to which particular results might apply, and how the mathematical results should be interpreted in practical situations. These conclusions can in turn be compared with what seem to be the important properties of the situation being modeled (such as the existence of bidder rings at auctions, for example). Thus the question of which are the essential features of a situation can be addressed not merely by empirical observation, but also by analysis of related models. The process of weighing and interpreting the significance of different mathematical assumptions and results is an essential part of both the modeling and analysis. This fundamental relationship among observation, modeling, and analysis is part of what we have sought to illustrate.

As we noted in the introduction, not only have we analyzed a variety of models, we have employed some very different tools in the analysis. The chief analytical tools, stability and strategic equilibrium, are a mix drawn from what has traditionally been called cooperative and noncooperative game theory. Their use here together should help make clear why this is not always a useful distinction, since these two approaches to game theory are complements rather than substitutes. Whereas we need to identify the rules of the game in great detail in order to speak about equilibrium, we can discuss the stability of a matching somewhat independently of the specific rules of the market. And one of the phenomena that emerges from these studies is that when the outcomes are unstable, agents have incentives to *change* the rules of the game, as when they decide to introduce a centralized matching procedure, or to defect from one.

Finally, what conclusions can we draw from the empirical observations we have so far been able to make of two-sided matching markets? Although some of this has been widely interpreted as evidence that "game theory works," our own view is that a somewhat more cautious interpretation is called for. First, there is now an enormous and varied body of game-theoretic work concerning a diversity of environments, but there has so far been very much less empirical work that can be thought of as providing tests of game-theoretic predictions. This is no doubt due to the difficulty of gathering the kind of detailed information about institutions and agents that game-theoretic theories employ, and for this reason much of the most interesting empirical work – pace Bob Aumann, who notes his opinion to the contrary in his gracious foreword to this book – has involved the relatively new practice of conducting controlled experiments under laboratory conditions. What has made the empirical work on two-sided matching markets different is that it has proved possible to identify markets for which the necessary information can be found. Which brings us to the question: How does the theory fare when tested on the markets observed to date?

Even here, the answer is a little complex. We certainly can't claim that the evidence supports the simple hypothesis that the outcome of two-sided matching markets will always be stable, since we have observed markets that employ unstable procedures and produce unstable matchings at least some of the time. And even those markets that eventually developed procedures to produce stable matchings – the American medical market, and those operated out of Edinburgh and Cardiff – operated for many years without such procedures before the problems they encountered in doing so led them to develop the rules they successfully use today.

The evidence is much clearer however, when we turn from simple predictions to conditional predictions. The available evidence strongly supports the hypothesis that if matching markets are organized in ways that produce unstable matchings, then they are prone to a variety of related problems and market failures that can largely be avoided if the markets are organized in ways that produce stable matchings.

Thus the empirical evidence very clearly suggests that game-theoretic considerations, concerning the incentives that different ways of organizing a market give to the agents, play an important role in determining how those markets will behave. It is this close observed connection between individual incentives and market behavior that suggests that, however game theory may need to be further developed as a descriptive theory, it has a critical role to play in helping us to understand and design the institutions through which we interact.

CHAPTER 10

Open questions and research directions

We close with a few open questions and suggestions of possible directions for further research. In keeping with the emphasis of the book on both analysis and modeling, some of the questions and directions are of each type; that is, some call for the statement and proof of theorems, whereas progress on others will (first) involve the construction of new models.

1. Since many entry level labor markets and other two-sided matching situations don't employ centralized matching procedures, and yet aren't observed to experience the kinds of market failure that seem to be associated with unstable matching, we can conjecture that at least some of these markets reach stable outcomes by means of decentralized decision making. So one of the chief modeling problems that will arise in studying such markets will be to develop decentralized models of stable matching. (We noted that a consequence of Theorem 2.33 is that a random process that begins from an arbitrary matching and continues by satisfying a randomly selected blocking pair must eventually converge with probability one to a stable matching, provided each blocking pair has a probability of being selected that is bounded away from zero. Perhaps this kind of result will provide the building blocks for models of decentralized matching.)

2. In Section 3.2.2 we considered how fast the number of stable matchings might grow as a function of the number of men and women in the market. We found that it might grow exponentially. This was a *worst case* analysis, which proceeded by showing that examples existed that exhibited exponential growth. For many practical purposes, it might be at least as useful to know the *average* size of the set of stable matchings as a function of the number of men and women (or firms and workers in many-to-one matching models). That is, for some assumptions about probability distributions over preferences (e.g., suppose most simply that all agents have independent preferences that give equal probability to

each permutation of possible mates), what is the mean and variance of the number of stable matchings as a function of the numbers of agents?

3. In Chapter 4 we saw that equilibrium strategies for the complete information model made heavy use of information about other players' preferences, but that far fewer conclusions could be drawn about equilibria in games where the information that players possessed about one anothers' preferences could be completely general. What kind of information is needed to compute an equilibrium strategy? What are the informational and computational requirements for determining if it is possible for a particular agent to manipulate in a given setting? (Such questions have not as yet drawn much attention, but Bartholdi, Tovey, and Trick 1989 investigate the computational requirements of manipulation in a voting context.)

4. In models of labor markets with two-career households, find reasonable assumptions about the preferences of married couples that assure the nonemptiness of the core (cf. Theorem 5.11). Under what conditions are pairwise stable matchings always in the core of such a market? Find an algorithm for identifying stable and core outcomes.

5. Identify an appropriate "refinement" (i.e., a subset) of Nash equilibrium misrepresentations that permits a sharper result than Theorem 5.18, perhaps along the lines of Theorem 4.16. Consider the nature of equilibrium misrepresentation with plausible assumptions about how much information agents can have about one another's preferences.

6. Find weaker conditions than substitutability that guarantee the nonemptiness of the core in many-to-one matching with complex preferences. Find necessary and sufficient conditions for pairwise stable outcomes to be in the core in models of many-to-one and many-to-many matching (recall Examples 6.9 and 5.24). What are the properties of the core in many-to-many matching?

7. In the marriage model, optimal stable matchings for the two sides of the market exist when preferences are strict. In the models of Part III, optimal stable matchings exist when preferences are not strict. In view of the fact that the results for the models of one-to-one matching with money closely parallel those for the marriage model, it seems likely that the precise similarities and differences between the two classes of models can be fully characterized, but this remains to be persuasively done. Perhaps the linear programming formulations of the two kinds of problems (in Chapter 8 and Section 3.2.4) can be used to bridge the gap. (Note added in proof: We have recently been able to show that in this connection the special role of strict preferences in the marriage model may be that they rule out "weakly blocking" pairs in which only one member of some pair strictly prefers a new matching: Such pairs cause no problem

in a market with continuously divisible money, since they can always be converted to conventional blocking pairs by a small transfer of funds.)

8. As mentioned in the guide to the literature following Chapters 2 and 8, highly structured models of the kind explored by Becker (1981), which make strong assumptions about the nature of agents' preferences, allow one to investigate how agents sort themselves out, that is, who is matched to whom. The models explored in this volume make few assumptions about preferences, but reveal a great deal of structure to the set of stable outcomes, some of which allow welfare comparisons (e.g., there are optimal stable outcomes for each side of the market). It seems likely that these two kinds of investigations can be profitably combined.

9. In the guide to the literature following Chapter 4, we mentioned the job search literature, which models agents' information before they know enough to be sure of their own preferences. As remarked earlier, this is the kind of model needed to study the problem facing, say, graduating medical students, *before* they have gone on interviews. However the strategic problems they face in deciding which interviews to go on (when the number they can go on is limited by the time available) are obviously related to the situation that will exist after they have formed their preferences from these interviews, which is the situation we have studied here. It therefore seems likely that these two kinds of models also can be profitably combined.

10. Sociologists consider a problem they call the "marriage squeeze," which can be (much too briefly) described as follows. Suppose men like to marry women who are about two years younger then themselves, and women like to marry men about two years older. Now consider what happens as a baby boom passes through the population. Men born early in the baby boom, or shortly before it, find that the cohort of women two years younger is larger than their own cohort, so the competition among men for the most desirable marriage partners (in terms of age) is not great. That is, women born early in the baby boom find that there are too few slightly older men. But after the peak of the boom, as the size of each cohort declines, the situation is reversed: Men born on the downside of the boom find that there are two few slightly younger women. To explore the different pattern of marriages in the populations before and after the peak of the boom, a model would be needed that combines some of the features of those suggested in the previous two remarks. That is, we would need to model both what kinds of preferences agents have, and what kind of information they have about their future prospects, when considering the possibility of a particular marriage. (Regarding preferences, the situation in which the second choice of both men and women is a three-year age difference in mates will be different from

the one in which this is the second choice of men, but women prefer a one-year age difference if they cannot find a mate two years older.) Such a model could be used to investigate which men marry which women in various situations, and perhaps more ambitious questions such as how the population makeup influences the age of marriage (e.g., by influencing the optimal length of search).

11. Finally, there will be many empirical questions that arise in investigating particular markets and two-sided matching situations. In Section 5.5 we remarked that it has been particularly convenient to begin the empirical investigation with markets that have some well-defined centralized matching procedure, but that it should be possible to study quite a wide variety of markets using the theoretical framework developed here. In particular, different institutional arrangements for matching agents to one another will offer new possibilities to test this kind of theory. The authors of this volume would be glad to learn the details of any matching procedures with which readers might be acquainted.

Bibliography

Numbers set in braces following a citation refer to the chapter(s) in which it is cited.

Alkan, Ahmet. 1986. Nonexistence of stable threesome matchings. *Mathematical Social Sciences*, **16**, 207–9. {2}

1988a. Existence and computation of matching equilibria. Bogazici University. Mimeo. {9}

1988b. Auctioning several objects simultaneously. Bogazici University. Mimeo. {9}

Alkan, Ahmet, and David Gale. 1988. The core of the matching game. *Games and Economic Behavior*, 1990, **2**, 203–12. {9}

Allison, Lloyd. 1983. Stable marriages by coroutines. *Information Processing Letters*, **16**, 61–5. {3}

Arrow, Kenneth J. 1951. *Social Choice and Individual Values*. New York: Wiley, 2d ed. 1963, Cowles Foundation Monograph, Yale University Press. {4}

Ashenfelter, Orley. 1989. How auctions work. *Journal of Economic Perspectives*, **3**, 23–36. {7}

Aumann, Robert J. 1964. Markets with a continuum of traders. *Econometrica*, **32**, 39–50. {3}

Balinski, M. L., and David Gale. 1987. On the core of the assignment game. In *Functional Analysis, Optimization, and Mathematical Economics: A Collection of Papers Dedicated to the Memory of Leonid Vital'evich Kantorovich*, Lev J. Leifman (ed.). Oxford: Oxford University Press, 1990, 274–89. {8}

Bartholdi, John J., III, and Michael A. Trick. 1986. Stable matching with preferences derived from a psychological model. *Operations Research Letters*, **5**, 165–9. {2}

Bartholdi, John J., III, Craig A. Tovey, and Michael A. Trick. 1989. The computational difficulty of manipulating an election. *Social Choice and Welfare*, **6**, 227–41. {10}

Becker, Gary S. 1981. *A Treatise on the Family*. Cambridge: Harvard University Press. {2, 8, 10}

Bennett, Elaine. 1988. Consistent bargaining conjectures in marriage and matching. *Journal of Economic Theory*, **45**, 392–407. {8}

Bergstrom, Theodore, and Richard Manning. 1982. Can courtship be cheatproof? Personal communication. {4}

Bird, Charles G. 1984. Group incentive compatibility in a market with indivisible goods. *Economics Letters*, **14**, 309–13. {4}

Birkhoff, Garrett. 1973. *Lattice Theory.* 3d ed. Vol 25 of *American Mathematical Society Colloquium Publications.* Providence: American Mathematical Society. {3}

Blair, Charles. 1984. Every finite distributive lattice is a set of stable matchings. *Journal of Combinatorial Theory* (Series A), 37, 353-6. {3}

1988. The lattice structure of the set of stable matchings with multiple partners. *Mathematics of Operations Research,* 13, 619-28. {6}

Brams, Steven J., and Philip D. Straffin, Jr. 1979. Prisoners' dilemma and professional sports drafts. *American Mathematical Monthly,* 86, 80-8. {4}

Brissenden, T. H. F. 1974. Some derivations from the marriage bureau problem. *The Mathematical Gazette,* 58, 250-7. {3}

Cassady, Ralph, Jr. 1967. *Auctions and Auctioneering.* Berkeley: University of California Press. {1}

Checker, Armand. 1973. The national intern and resident matching program, 1966-72. *Journal of Medical Education,* 48, 106-9. {1}

Crawford, Vincent P. 1988. Comparative statics in matching markets. *Journal of Economic Theory,* 54, 1991, 389-400. {6}

Crawford, Vincent P., and Elsie Marie Knoer. 1981. Job matching with heterogeneous firms and workers. *Econometrica,* 49, 437-50. {6, 8, 9}

Crawford, Vincent P., and Sharon C. Rochford. 1986. Bargaining and competition in matching markets. *International Economic Review,* 27, 329-48. {8}

Cremer, Jacques, and Richard P. McLean. 1985. Optimal selling strategies under uncertainty for a discriminating monopolist when demands are interdependent. *Econometrica,* 53, 345-61. {7}

Curiel, Imma J. 1988. Cooperative game theory and applications. Ph.D. diss., Katholieke Universiteit van Nijmegen. {8}

Curiel, Imma J., and Stef H. Tijs. 1985. Assignment games and permutation games. *Methods of Operations Research,* 54, 323-34. {8}

Dantzig, George B. 1963. *Linear Programming and Extensions.* Princeton: Princeton University Press. {8}

Dasgupta, Partha, Peter Hammond, and Eric Maskin. 1979. The implementation of social choice rules: some general results on incentive compatibility. *Review of Economic Studies,* 46, 185-216. {4}

Debreu, Gerard, and Herbert Scarf. 1963. A limit theorem on the core of an economy. *International Economic Review,* 4, 235-46. {3}

Demange, Gabrielle. 1982. Strategyproofness in the assignment market game. Laboratoire d'Econometrie de l'Ecole Polytechnique, Paris. Mimeo. {8}

1987. Nonmanipulable cores. *Econometrica,* 55, 1057-74. {3}

Demange, Gabrielle, and David Gale. 1985. The strategy structure of two-sided matching markets. *Econometrica,* 53, 873-88. {2, 9}

Demange, Gabrielle, David Gale, and Marilda Sotomayor. 1986. Multi-item auctions. *Journal of Political Economy,* 94, 863-72. {8}

1987. A further note on the stable matching problem. *Discrete Applied Mathematics,* 16, 217-22. {3, 4, 8}

Diamond, Peter, and Eric Maskin. 1979. An equilibrium analysis of search and breach of contract, I: steady states. *Bell Journal of Economics,* 10, 282-316. {4}

1982. An equilibrium analysis of search and breach of contract, II: a non-steady state example. *Journal of Economic Theory,* 25, 165-95. {4}

Dubins, L. E., and D. A. Freedman. 1981. Machiavelli and the Gale–Shapley algorithm. *American Mathematical Monthly,* **88,** 485–94. {4}

Edgeworth, F. Y. 1881. *Mathematical Psychics: An Essay on the Application of Mathematics to the Moral Sciences.* London: Kegan Paul. {3}

Feder, Tomás. 1989. A new fixed point approach for stable networks and stable marriages. *Proceedings of the Twenty-first Annual ACM Symposium on Theory of Computing,* 513–22. {3}

Francis, N. D., and D. I. Fleming. 1985. Optimum allocation of places to students in a national university system. *BIT,* **25,** 307–17. {3}

Gale, David. 1960. *The Theory of Linear Economic Models.* New York: McGraw Hill. {8}

1968. Optimal assignments in an ordered set: an application of matroid theory. *Journal of Combinatorial Theory,* **4,** 176–80. {3}

1984. Equilibrium in a discrete exchange economy with money. *International Journal of Game Theory,* **13,** 61–4. {9}

Gale, David, and Lloyd Shapley. 1962. College admissions and the stability of marriage. *American Mathematical Monthly,* **69,** 9–15. {2, 5}

Gale, David, and Marilda Sotomayor. 1985a. Some remarks on the stable matching problem. *Discrete Applied Mathematics,* **11,** 223–32. {2, 3, 4, 5}

1985b. Ms Machiavelli and the stable matching problem. *American Mathematical Monthly,* **92,** 261–8. {2, 4}

Gardenfors, Peter. 1973. Assignment problem based on ordinal preferences. *Management Science,* **20,** 331–40. {3}

1975. Match making: assignments based on bilateral preferences. *Behavioral Science,* **20,** 166–73. {2}

Garey, M. R., and D. S. Johnson. 1979. *Computers and Intractability.* Freeman: San Francisco. {3, 5}

Gibbard, Alan. 1973. Manipulation of voting schemes: a general result. *Econometrica,* **41,** 587–601. {4}

Gillies, D. B. 1953a. Some theorems on *N*-person games. Ph.D. diss., Princeton University. {3, 5}

1953b. Locations of solutions. In *Report of an Informal Conference on the Theory of N-Person Games,* Princeton University. {3, 5}

Graham, Daniel A., and Robert C. Marshall. 1984. Bidder coalitions at auctions. Duke University Department of Economics. Mimeo. {1, 7}

1987. Collusive bidder behavior at single-object second-price and English auctions. *Journal of Political Economy,* **95,** 1217–39. {1, 7}

Graham, Daniel A., Robert C. Marshall, and Jean-Francois Richard. 1987a. Auctioneer's behavior at a single object English auction with heterogeneous non-cooperative bidders. Working paper no. 87-01, Duke University Institute of Statistics and Decision Sciences. {7}

Graham, Daniel A., Robert C. Marshall, and Jean-Francois Richard. 1987b. Differential payments within a bidder coalition and the Shapley value. *American Economic Review,* **80,** 1990, 493–510. {7}

Granot, Daniel. 1984. A note on the room-mates problem and a related revenue allocation problem. *Management Science,* **30,** 633–43. {2}

Green, Jerry R., and Jean-Jacques Laffont. 1979. Incentives in public decision-making. Amsterdam: North-Holland. {4, 5}

Gusfield, Dan. 1987. Three fast algorithms for four problems in stable marriage. *SIAM Journal of Computing,* **16,** 111–28. {3}

1988. The structure of the stable roommate problem: efficient representation and enumeration of all stable assignments. *SIAM Journal on Computing,* **17**, 742–69. {2}

Gusfield, Dan, and Robert W. Irving. 1989. *The Stable Marriage Problem: Structure and Algorithms.* Cambridge: MIT Press. {3}

Gusfield, Dan, Robert W. Irving, Paul Leather, and M. Saks. 1987. Every finite distributive lattice is a set of stable matchings for a *small* stable marriage instance. *Journal of Combinatorial Theory A,* **44**, 304–9. {3}

Hall, P. 1935. On representatives of subsets. *Journal of the London Mathematical Society,* **10**, 26–30. {8}

Harrison, Glenn W., and Kevin A. McCabe. 1989. Stability and preference distortion in resource matching: an experimental study of the marriage market. Department of Economics, University of New Mexico. Mimeo. {5}

Harsanyi, John C. 1967. Games with incomplete information played by "Bayesian" players, I: the basic model. *Management Science,* **14**, 159–82. {4}

1968a. Games with incomplete information played by "Bayesian" players, II: Bayesian equilibrium points. *Management Science,* **14**, 320–34. {4}

1968b. Games with incomplete information played by "Bayesian" players, III: the basic probability distribution of the game. *Management Science,* **14**, 486–502. {4}

Holmstrom, Bengt, and Roger Myerson. 1983. Efficient and durable decision rules with incomplete information. *Econometrica,* **51**, 1799–1819. {7}

Hull, M. Elizabeth C. 1984. A parallel view of stable marriages. *Information Processing Letters,* **18**, 63–6. {3}

Hwang, J. S. 1978. Complete unisexual stable marriages. *Soochow Journal of Mathematics,* **4**, 149–51. {2}

1986. The algebra of stable marriages. *International Journal of Computer Mathematics,* **20**, 227–43. {2}

n.d. Modelling on college admissions in terms of stable marriages. Academia Sinica. Mimeo. {3, 4}

Hwang, J. S., and H. J. Shyr. 1977. Complete stable marriages. *Soochow Journal of Mathematical and Natural Sciences,* **3**, 41–51. {2}

Hylland, Aanund, and Richard Zeckhauser. 1979. The efficient allocation of individuals to positions. *Journal of Political Economy,* **87**, 293–314. {3}

Irving, Robert W. 1985. An efficient algorithm for the stable room-mates problem. *Journal of Algorithms,* **6**, 577–95. {3}

1986. On the stable room-mates problem. Department of Computing Science, University of Glasgow. Mimeo. {2}

Irving, Robert W., and Paul Leather. 1986. The complexity of counting stable marriages. *SIAM Journal of Computing,* **15**, 655–67. {3}

Irving, Robert W., Paul Leather, and Dan Gusfield. 1987. An efficient algorithm for the "optimal" stable marriage. *Journal of the ACM,* **34**, 532–43. {3}

Itoga, Stephen Y. 1978. The upper bound for the stable marriage problem. *Journal of the Operational Research Society,* **29**, 811–14. {3}

1981. A generation of the stable marriage problem. *Journal of the Operational Research Society,* **32**, 1069–74. {3}

1983. A probabilistic version of the stable marriage problem. *BIT,* **23**, 161–9. {3}

Jones, Philip C. 1983. A polynomial time market mechanism. *Journal of Information and Optimization Sciences,* **4**, 193–203. {6}

Kalai, Ehud, and Dov Samet. 1985. Are Bayesian–Nash incentives and implementations perfect? MEDS Department, Northwestern University. Mimeo. {4}

Kamecke, Ulrich. 1987. A generalization of the Gale–Shapley algorithm for monogamous stable matchings to the case of continuous transfers. Discussion paper, Rheinische Friedrich-Wilhelms Universitat, Bonn. [8]

1989. Non-cooperative matching games. *International Journal of Game Theory*, **18**, 423–31.

Kaneko, Mamoru. 1976. On the core and competitive equilibria of a market with indivisible goods. *Naval Research Logistics Quarterly*, **23**, 321–37. [2, 8]

1982. The central assignment game and the assignment markets. *Journal of Mathematical Economics*, **10**, 205–32. [2, 8]

1983. Housing markets with indivisibilities. *Journal of Urban Economics*, **13**, 22–50. [2]

Kaneko, Mamoru, and Myrna Holtz Wooders. 1982. Cores of partitioning games. *Mathematical Social Sciences*, **3**, 313–27. [6, 8]

1985. The core of a game with a continuum of players and finite coalitions: nonemptiness with bounded sizes of coalitions. Institute for Mathematics and its Applications, University of Minnesota. Mimeo. [3]

1986. The core of a game with a continuum of players and finite coalitions: the model and some results. *Mathematical Social Sciences*, **12**, 105–37. [3]

Kaneko, Mamoru, and Yoshitsugu Yamamoto. 1986. The existence and computation of competitive equilibria in markets with an indivisible commodity. *Journal of Economic Theory*, **38**, 118–36. [8]

Kapur, Deepak, and Mukkai S. Krishnamoorthy. 1985. Worst-case choice for the stable marriage problem. *Information Processing Letters*, **21**, 27–30. [3]

Kelso, Alexander S., Jr., and Vincent P. Crawford. 1982. Job matching, coalition formation, and gross substitutes. *Econometrica*, **50**, 1483–1504. [2, 6, 8]

Knuth, Donald E. 1976. *Marriages Stables*. Montreal: Les Presses de l'Université de Montreal. [2, 3]

Leonard, Herman B. 1983. Elicitation of honest preferences for the assignment of individuals to positions. *Journal of Political Economy*, **91**, 461–79. [8]

Masarani, F., and S. S. Gokturk. 1988. On the probabilities of the mutual agreement match. *Journal of Economic Theory*, **44**, 192–201. [2]

McVitie, D. G., and L. B. Wilson. 1970a. Stable marriage assignments for unequal sets. *BIT*, **10**, 295–309. [2]

1970b. The application of the stable marriage assignment to university admissions. *Operational Research Quarterly*, **21**, 425–33. [3]

1971. The stable marriage problem. *Communications of the ACM*, **14**, 486–92. [3]

Milgrom, Paul R., and Robert J. Weber. 1982. A theory of auctions and competitive bidding. *Econometrica*, **50**, 1089–122. [7]

Mo, Jie-ping. 1988a. Entry and structures of interest groups in assignment games. *Journal of Economic Theory*, **46**, 66–96. [2, 8]

1988b. Global stability analysis of assignment games. Institute of Economics, Academia Sinica. Mimeo. [8]

Mo, Jie-ping, and Jyh-chi Gong. 1989. Shapley values and second differentials in the entry problem of game theory. Academia Sinica. Mimeo. [8]

Mo, Jie-ping, Pei-sung Tsai, and Sheng-chang Lin. 1988. Pure and minimal overdemanded sets: a note on Demange, Gale, and Sotomayor. Institute of Economics, Academia Sinica. Mimeo. [8]

Moldovanu, Benny. 1988. Stable bargained equilibria for assignment games without side payments. *International Journal of Game Theory*, **19**, 1990, 171–91. [8]

Mongell, Susan J. 1988. Sorority rush as a two-sided matching mechanism: a game-theoretic analysis. Ph.D. diss., Department of Economics, University of Pittsburgh. [5]

Mongell, Susan J., and Alvin E. Roth. 1986. A note on job matching with budget constraints. *Economics Letters,* **21,** 135-8. [6]

    1989. Sorority rush as a two-sided matching mechanism. *American Economic Review,* **81,** 1991, 441-64. [5]

Mortensen, Dale T. 1982. The matching process as a noncooperative bargaining game. In *The Economics of Information and Uncertainty,* J. McCall (ed.). Chicago: University of Chicago Press, 233-58. [4]

Moulin, Herve. 1986. *Game Theory for the Social Sciences.* 2d ed. New York: New York University Press. [4]

Myerson, Roger B. 1981. Optimal auction design. *Mathematics of Operations Research,* **6,** 58-73. [7]

    1983. The basic theory of optimal auctions. In *Auctions, Bidding and Contracting: Uses and Theory,* R. Englebrecht-Wiggans, M. Shubik, and R. Stark (eds.). New York: New York University Press. [7]

    1985. Bayesian equilibrium and incentive-compatibility: an introduction. In *Social Goals and Social Organizations: Essays in Memory of Elisha Pazner,* L. Hurwicz, D. Schmeidler, and H. Sonnenschein (eds.). Cambridge: Cambridge University Press. [4]

Nash, John F. Jr. 1951. Noncooperative games. *Annals of Mathematics,* **54,** 286-95. [4]

NIRMP Directory. 1979. Evanston, IL: National Resident Matching Program. [5]

NRMP Directory. 1987. Evanston, IL: National Resident Matching Program. [5]

Owen, Guillermo. 1975. On the core of linear production games. *Mathematical Programming,* **9,** 358-70. [8]

Peleg, Bezalel. 1978. Consistent voting systems. *Econometrica,* **46,** 153-70. [4]

    1984. *Game Theoretic Analysis of Voting in Committees.* Econometric Society Monographs. New York: Cambridge University Press. [4]

    1988. Axiomatizations of the core. In *Handbook of Game Theory,* R. J. Aumann and S. Hart (eds.). Forthcoming. [3]

Prasad, Kislaya. 1987. The complexity of games II: assignment games and indices of power. Department of Economics, Syracuse University. Mimeo. [6]

Proll, L. G. 1972. A simple method of assigning projects to students. *Operational Research Quarterly,* **23,** 195-201. [3]

Quinn, Michael J. 1985. A note on two parallel algorithms to solve the stable marriage problem, *BIT,* **25,** 473-6. [3]

Quint, Thomas. 1987a. Elongation of the core in an assignment game. Technical report, IMSSS, Stanford University. [8]

    1987b. A proof of the nonemptiness of the core of two-sided matching markets. CAM report no. 87-29, Department of Mathematics, UCLA. [8]

    1988a. An algorithm to find a core point for a two-sided matching model. CAM report no. 88-03, Department of Mathematics, UCLA. [8]

    1988b. The core of an *m*-sided assignment game. *Games and Economic Behavior,* **3,** 1991, 487-503. [8]

Quinzii, Martine. 1984. Core and competitive equilibria with indivisibilities. *International Journal of Game Theory,* **13,** 41-60. [6, 9]

Rochford, Sharon C. 1984. Symmetrically pairwise-bargained allocations in an assignment market. *Journal of Economic Theory,* **34,** 262-81. [8]

Ronn, Eytan. 1986. On the complexity of stable matchings with and without ties. Ph.D. diss., Yale University. {5}

1987. NP-complete stable matching problems. Computer Science Department, Technion - Israel Institute of Technology. Mimeo. {5}

Roth, Alvin E. 1982a. The economics of matching: stability and incentives. *Mathematics of Operations Research*, **7**, 617-28. {2, 4}

1982b. Incentive compatibility in a market with indivisible goods. *Economics Letters*, **9**, 127-32. {4}

1984a. The evolution of the labor market for medical interns and residents: a case study in game theory. *Journal of Political Economy*, **92**, 991-1016. {1, 2, 4, 5}

1984b. Misrepresentation and stability in the marriage problem. *Journal of Economic Theory*, **34**, 383-7. {4}

1984c. Stability and polarization of interests in job matching. *Econometrica*, **52**, 47-57. {6}

1985a. The college admissions problem is not equivalent to the marriage problem. *Journal of Economic Theory*, **36**, 277-88. {1, 5}

1985b. Common and conflicting interests in two-sided matching markets. *European Economic Review*, **27**, 75-96. (Special issue on Market Competition, Conflict, and Collusion) {5}

1985c. Conflict and coincidence of interest in job matching: some new results and open questions. *Mathematics of Operations Research*, **10**, 379-89. {6}

1986. On the allocation of residents to rural hospitals: a general property of two-sided matching markets. *Econometrica*, **54**, 425-7. {1, 5}

(ed.). 1987. *Laboratory Experimentation in Economics: Six Points of View.* Cambridge: Cambridge University Press. {5}

1988a. Laboratory experimentation in economics: a methodological overview. *Economic Journal*, **98**, 974-1031. {5, 7}

1988b. *The Shapley Value: Essays in Honor of Lloyd S. Shapley.* Cambridge: Cambridge University Press. {8}

1989a. Two-sided matching with incomplete information about others' preferences. *Games and Economic Behavior*, **1**, 191-209. {4}

1989b. A natural experiment in the organization of entry level labor markets: regional markets for new physicians and surgeons in the U.K. *American Economic Review*, **81**, 1991, 415-40. {1, 5, 6}

Roth, Alvin E., and Andrew Postlewaite. 1977. Weak versus strong domination in a market with indivisible goods. *Journal of Mathematical Economics*, **4**, 131-7. {3}

Roth, Alvin E., Uriel G. Rothblum, and John H. Vande Vate. 1990. Stable matching and linear programming. *Mathematics of Operations Research*, forthcoming. {3}

Roth, Alvin E., and Marilda Sotomayor. 1988. Interior points in the core of two-sided matching problems. *Journal of Economic Theory*, **45**, 85-101. {8, 9}

1989. The college admissions problem revisited. *Econometrica*, **57**, 559-70. {5}

Roth, Alvin E., and John H. Vande Vate. 1989. Incentives in two-sided matching with random stable mechanisms. *Economic Theory*, **1**, 1991, 31-44. {4}

Roth, Alvin E., and John H. Vande Vate. 1990. Random paths to stability in two-sided matching. *Econometrica*, **58**, 1990, 1475-80. {2}

Rothblum, Uriel G. 1989. Characterization of stable matchings as extreme points of a polytope. *Mathematical Programming*, forthcoming. {3}

Samet, Dov, and Eitan Zemel. 1984. On the core and dual set of linear programming games. *Mathematics of Operations Research,* **9,** 309-16. {8}

Sasaki, Hiroo. 1988. Axiomatization of the core for two-sided matching problems. Economics discussion paper no. 86, Faculty of Economics, Nagoya City University, Nagoya, Japan. {3}

Sasaki, Hiroo, and Manabu Toda. 1986. Marriage problem reconsidered: externalities and stability. Department of Economics, University of Rochester. Mimeo. {6}

Satterthwaite, Mark A. 1975. Strategy-proofness and Arrow's conditions: existence and correspondence theorems for voting procedures and social welfare functions. *Journal of Economic Theory,* **10,** 187-217. {4}

Scotchmer, Suzanne, and Myrna Holtz Wooders. 1989. Monotonicity in games that exhaust gains to scale. University of California, Berkeley. Mimeo. {2}

Shapley, Lloyd S. 1953a. Open questions. In *Report of an Informal Conference on the Theory of N-Person Games,* Princeton University. {3, 5}

1953b. A value for *n*-person games. In *Contributions to the Theory of Games II,* H. W. Kuhn and A. W. Tucker (eds.). Annals of Mathematics Studies No. 28, Princeton: Princeton University Press, pp. 307-17. {8}

1962. Complements and substitutes in the optimal assignment problem. *Naval Research Logistics Quarterly,* **9,** 45-8. {2, 8}

Shapley, Lloyd S., and Herbert Scarf. 1974. On cores and indivisibility. *Journal of Mathematical Economics,* **1,** 23-8. {3}

Shapley, Lloyd S., and Martin Shubik. 1972. The assignment game I: the core. *International Journal of Game Theory,* **1,** 111-30. {2, 6, 8}

Shubik, Martin. 1959. Edgeworth market games. In *Contributions to the Theory of Games,* Vol. 4, R. D. Luce and A. W. Tucker (eds.). Princeton: Princeton University Press. {3}

1982. *Game Theory in the Social Sciences: Concepts and Solutions.* Cambridge: MIT Press. {3}

1984. *A Game Theoretic Approach to Political Economy.* Cambridge: MIT Press. {8}

Sondak, Harris, and Max H. Bazerman. 1988. Matching and negotiation processes in quasi-markets. *Organizational Behavior and Human Decision Processes,* forthcoming. {5}

Sotomayor, Marilda. 1986a. On incentives in a two-sided matching market. Working paper, Department of Mathematics, Pontificia Universidade Catolica do Rio de Janeiro. {9}

1986b. The simple assignment game versus a multiple assignment game. Working paper, Department of Mathematics, Pontificia Universidade Catolica do Rio de Janeiro. {8}

1987. Further results on the core of the generalized assignment game. Working paper, Department of Mathematics, Pontificia Universidade Catolica do Rio de Janeiro. {9}

1988. The multiple partners game. *Equilibrium and Dynamics: Essays in Honor of David Gale,* William Brock and Mukul Majumdar (eds.). In preparation. {8}

Stalnaker, John M. 1953. The matching program for intern placement: the second year of operation. *Journal of Medical Education,* **28,** 13-19. {1, 5}

Subramanian, Ashok. 1989. A new approach to stable matching problems. Stanford University. Mimeo. {3}

Thompson, Gerald L. 1980. Computing the core of a market game. *Extremal Methods and Systems Analysis*, A. V. Fiacco and K. O. Kortanek (eds.). Lecture Notes in Economics and Mathematical Systems no. 174, Berlin: Springer, pp. 312-34. {8}

Thomson, William. 1986. Reversal of asymmetries of allocation mechanisms under manipulation. *Economics Letters*, **21**, 227-30. {7}

Toda, Manabu. 1988. The consistency of solutions for marriage problems. Department of Economics, University of Rochester. Mimeo. {3}

Tseng, S. S., and R. C. T. Lee. 1984. A parallel algorithm to solve the stable marriage problem. *BIT*, **24**, 308-316. {3}

Vande Vate, John H. 1989. Linear programming brings marital bliss. *Operations Research Letters*, **8**, 147-53. {3}

Vickrey, W. 1961. Counterspeculation, auctions, and competitive sealed tenders. *Journal of Finance*, **16**, 8-37. {7}

von Neumann, John, and Oskar Morgenstern. 1944. *Theory of Games and Economic Behavior*. Princeton: Princeton University Press. {3, 7}

Wilson, L. B. 1972. An analysis of the stable marriage assignment algorithm. *BIT*, **12**, 569-75. {3}

1977. Assignment using choice lists. *Operational Research Quarterly*, **28**, 569-78. {3}

Wood, Robert O. 1984. A note on incentives in the college admissions market. Stanford University. Mimeo. {5}

Name index

Alkan, 24, 51, 225, 239
Allison, 76
Anderson, 151
Arrow, 121
Ashenfelter, 201
Aumann, xi, 76, 243

Balinski, 220
Bartholdi, 51, 246
Bazerman, 169
Becker, 51, 220, 247
Bennett, 220
Bergstrom, 119
Bird, 120
Birkhoff, 71, 75
Blair, 60, 75, 177, 185
Brams, 120
Brissenden, 76

Cassady, 9, 13
Checker, 13
Conway, 36, 50, 60
Crawford, 50, 171, 177, 181, 182, 184,
    185, 219, 220, 225, 238
Cremer, 201
Curiel, 220

Dantzig, 204, 206, 219
Dasgupta, 121
Debreu, 75
Demange, 50, 56, 75, 93, 120, 211, 212,
    214, 219, 222, 226, 227, 228, 229,
    230, 231, 234, 236, 238

Diamond, 121
Dubins, 90, 92, 120

Edgeworth, 75

Feder, 77
Fleming, 75
Francis, 75
Freedman, 90, 92, 120

Gale, 23, 27, 32, 41, 43, 44, 47, 50, 56,
    75, 93, 96, 100, 102, 106, 120, 165,
    166, 168, 169, 170, 211, 219, 220,
    222, 225, 226, 227, 228, 229, 230,
    231, 232, 233, 234, 236, 238, 239
Gardenfors, 51, 75
Garey, 76, 169
Gibbard, 121
Gillies, 75, 170
Gokturk, 51
Gong, 219
Graham, 9, 13, 199, 201
Granot, 51
Green, 121, 170
Gusfield, 51, 68, 75, 76

Hammond, 121
Harrison, 149
Harsanyi, 109, 120
Holmstrom, 201
Hull, 76
Hwang, 51, 56, 75, 120
Hylland, 75

259

Name index

Subject index